ANTHOLOGIN

Guido Tommasi Editore

Contents

Preface – 7

Introduction – 10

Chapter I – From Salerno to the world – 13

Chapter II – The world is gin – 35

Chapter III – Not by juniper alone... – 69

Chapter IV – The art of distillation – 97

Chapter V – Gin and its rules – 115

Chapter VI – The Gin and Tonic and other stories – 127

Chapter VII – 100 not-to-miss gins – 169

Index – 283

Bibliography – 287

Acknowledgments – 288

Preface

by Leonardo Leuci

I became a barman by chance, like many young Italian men do, to fund my holidays and buy myself a used scooter to take foreign girls around Rome on. And like so many stories, this one would later grow to become a great passion.
I was little more than sixteen years old, and yet I remember it like it was yesterday. A friend of a friend asked me to help out at a beach bar at the Litorale Romano nature reserve for the summer. It was a simple job. I'd leave early in the morning, passing the beach chairs, hot and iced coffees, the much hated orange juices, gelatos, cappuccinos. A classic routine, typical of a classic Italian bar.
The days flew by, perhaps because behind the bar, hours seemed to pass much more quickly. Often there was no time to even think. Yet sometimes, while serving a glass of water or whiskey, I'd catch a glimpse of the people all around me and realize that a bar, whether it's one in a grand hotel or some far-flung locale, is magical, unique. There's really no place like a bar. Thousands of lives, faces, eyes, and souls come together in a setting that exists solely to serve drinks. More than anything else, though, a bar unites people and can become a second home, if only for a few hours.
How many stories, loves, adventures, and friendships have started out in a bar?
And so it was there, in that little bar on the beach, that I met gin for the first time. It was one of those encounters that changes your life. Or at least it was for me.
A scorching hot August afternoon, and I had just finished work when the "boss" came over to ask me if I could do a night shift. Being one staff person short, the bar could use an extra pair of hands.
Though tired, I was excited at the prospect, because evenings were the world of grown-ups, a time when the bar transformed: the lights dimmed, the music turned up, the coffee cups and teaspoons disappeared to make

way for buckets of ice and tiny plates carrying orange and lemon slices ready to drop into drinks. I still laugh today when I think how improvised and erratic it all was, the job that to me seemed like the best job in the world at that time.

Before the evening began, the "boss" took me aside and began to explain how to make the most popular drinks. It was all new to me. Before that moment, I had at most poured a glass of sambuca or limoncello liqueur, but that night was different. That night, vodka and rum were the Queen and King of the bar. Rum and cokes and vodka lemons reigned. "We make a whisky and cola like a Cuba Libre but without lemon", I was told. "Someone might ask you for something strong. In which case, send them to me and I'll take care of it", said a colleague. He wasn't a barman: he worked at an airport all week, and on Fridays and Saturdays he earned a little extra this way. He seemed to know so much about bars. He could even make a Piña Colada. Back then I didn't even know what a Piña Colada was.

The shelves held just a few bottles: two or three types of blended whisky, a bottle of vodka, a bottle or two of amaro, two bottles of rum, a couple Italian brandies, a cheap cognac, a bottle of coconut liqueur, and little else. Though all very disorganized, to me it was like one of those bars I'd seen only in the movies.

Among the various bottles, one in particular stood out above the rest. It looked different, like a bottle from another era: simple and elegant, not seeking notice. Upon closer look, it was clear this bottle had a story to tell. I will never forget the response I got from the "boss" when I asked him what gin was used for. "Gin is for people who understand us", he said. "People who know how to drink: You make them a Gin and Tonic, a Martini Cocktail and a White Lady". That's what he said. Today I realize how naive that was, though at that moment it was like discovering a new world, and a different bar, one previously unknown to me but that I would soon come to love.

Over time, and with the rise of the flair bartending phenomenon, many spirits changed their personas, altering and bending to trends to become plain and often generic drink ingredients. Gin has not, however. For better or worse, gin remains faithful to its history: in cocktails, it's always in the lead role, never a simple cameo appearance. Because, as my first "boss" ingeniously put it more than twenty-five years ago: "Those who know how to drink, drink gin". Think about it: perhaps this will seem trivial, but in the bar world there is nothing more iconic than the Gin Martini. Not even the constant hammering taken for over half a century by the spy with a license to kill could tarnish gin's mythical status. Glorious gin continues to nourish the passion of millions of bartenders. Drinkers, too, those who take a seat at the bar and, when ordering, choose to be original rather than mimic some fictional character with his "shaken not stirred" vodka.

So why is gin different? In essence, gin is a distilled spirit like countless others, but what makes it unique?

While I don't believe this will hold true for everyone, I have dedicated several years to finding out, to truly understanding gin. It's not something easily explained. Simply put, one day you find yourself making a Gin Martini, with a mother's love and a surgeon's precision, and out of the blue it hits you: that nothing else will ever bring as much satisfaction to a barman as a simple gin, iced to perfection and flavored slightly with a splash of French vermouth. And let's face it: the satisfaction in preparing this gin classic is matched only by the drinking of it (yet perhaps only after we have come to understand it).

The world has changed, yet gin continues to embody one of the most crucial elements in the realm of mixology, indeed perhaps the most important. Hundreds of new products have joined the classic brands. Gin has shed the historical connotations that bind it to an exclusively British world, having matured into a global product that represents so much—beauty, difference, nuances—all while remaining true to itself and welcoming a myriad of influences, delicate and diverse.

The rise of the Negroni, the bold reappearance of the Martinez and Tuxedo, and the return of the Martini Cocktail (a cocktail I like to label "democratic": no longer a cocktail of extremes, neither super dry nor super strong, this cocktail has finally become accessible to the many and not the few, thanks to the rediscovery of sweet, delicate cocktail recipes in the second half of the 1800s), and above all the remarkable rise of the Gin and Tonic, a cocktail claiming the great distinction of having relaunched the ailing long drink sector—all of this is proof that gin is more alive than ever, making its mark on this new Golden Age of global bartending, just as it has always done. Gin is always in the leading role, never the supporting one.

Leonardo Leuci
President of the Jerry Thomas Speakeasy – Rome

Introduction

by Maurizio Maestrelli

"The Gin and Tonic has saved more Englishmen's lives, and minds, than all the doctors in the Empire"

Winston Churchill

One afternoon some years ago, I found myself seated rather uncomfortably on a chair, attending what had been presented to me as a lesson on gin and its secrets, led by Samuele Ambrosi, who in turn was introduced as an internationally award winning barman with extensive experience, bar owner, and a trainer of rising generations of bartenders.
His was a stellar résumé, to say the least. The only problem was that my chair was truly uncomfortable. And honestly, I doubted that Ambrosi could solve this particular problem.
But he did.
An over three-hour-long monologue passed in a heartbeat, with anecdotes, some clichés dispelled, tastings of various gins, and many other first-person experiences. An exciting tale, delivered with enviable verve and brimming with ideas for someone like myself, someone always on the lookout for a good story to write. That occasion was the final confirmation that the fascination spirits hold goes well beyond aromas and flavor, beyond technical traits, production systems, and aging methods. Because behind all this are gin's stories, the historical events that gave rise to or influenced their evolution, the various political and economic influences, alongside the more circumstantial yet no less fascinating episodes involving famous writers, actors, politicians, and countless others. This "behind the scenes" knowledge,

present in every glass that our trusted barman pours for us, is what renders each sip of a gin cocktail so special, smooth, and harmonious.

Realizing this leads to an appreciation and better understanding of the rebirth of mixology and renewed interest in distilled spirits. The first is no longer seen simply as mere, albeit fascinating, technical ability, while the second is, in my view, an entirely expected consequence. Within this rebirth of the mixology arts, gin seems to be aware of and appreciate its own moment of absolute glory. Of course, it's not the first such moment in gin's long history. The favorite child of juniper has for centuries been humanity's travelling companion, something perhaps forgotten from time to time, though never fully forgotten. And not once throughout its enduring relationship with humans has gin known such a diversified expansion, fabulously rich in sensory potential, as the historical moment we are currently living.

What better moment then to at last put those three hours of "Abrosian monologue" down on the page, hours which, once transformed from intention to reality, have grown to become so much, from entire days consumed by conversation to pages and pages of notes and what felt like time-space leaps that in the early days were a kind of alcoholic "big bang", yet then gradually began to reveal the threads holding everything together over centuries: various customs, consumption habits, and the ever-expanding ways this distilled classic rewards those who to better know it.

It is for this reason that *Anthologin* resembles a Greek trireme: At the helm, Samuele Ambrosi showed the route to follow; yours truly gave everything to rowing the oars; and lastly (given that esthetics form part of philosophy, as Aristotle taught), the fine works of internationally active illustrator Serena Conti formed "the trireme's sail", the images and graphics accompanying the text.

Inspired by: Samuele Ambrosi
Written by: Maurizio Maestrelli
Illustrated by: Serena Conti

Chapter I

From Salerno to the World
Gin's voyage through space and time

"… why don't you try it, gin-and-water is the source of all my inspiration"
Lord Byron

Nature has the extraordinary ability to astound us, even in the smallest things. Take unicellular organisms, for example. We could consider them human beings, potentially. Accepting, of course, that Darwin was right. But let's take another example, the bluish berries protected by needle-like leaves, "perched" on a bush, capable of surviving both a dry summer and a cold winter, with very little water. In itself, this is already a kind of miracle. But add to this that the berries in question, which to be precise should be called "cones", grow only on the female plants and emit an intense, penetrating aroma. Indeed, all of this is found in a glass of gin.

The virtues of juniper

Before getting to that glass of gin, however, we must start with the juniper berry itself. And note, too, that there's juniper and there's juniper. Some are thirty meters in height, going by the scientific name *Juniperus virginiana*, but more commonly known as Virginian juniper or eastern red cedar. This juniper isn't useful for much beyond being an excellent source of firewood. Then we have *Juniperus rigida*, fascinating but also not very useful, used mostly as an ornamental plant in the Far East. Next comes *Juniperus sabina*, also beautiful, but poisonous. Which brings us to the protagonist of these pages: *Juniperus communis*, whose scientific name would suggest a modest plant yet in truth is the only one of the bunch truly deserving of the limelight. This juniper is the "little big mother" who has contributed so much to the history of mixed drinks and, to some extent, the history of humankind. Don't let its Latin name, likely attributable to Linnaeus, deceive you. While it does indicates juniper's early production time (from the Latin *juni*, or "June", and *perus*, "production"), it also refers to the plant's difficult harvesting, as evidenced by the Latin-Celtic etymology derived from the term's crasis *gen*, meaning "shrub" in Celtic, and *prus*, which is Latin for "severe" or "harsh". In any case, its healing properties were known to the Egyptians, as reported in the Ebers Papyrus, a medical papyrus from around 1550 B.C. that cites juniper as a remedy for jaundice. The Greeks later used it as a cure for stomach cramps, while the Romans steeped it in wine as a general treatment for various ailments. In his *Naturalis Historia*, Pliny the Elder mentions it some twenty-two times, praising its diuretic, antioxidant, and, when steeped in wine for long periods, astringent properties.

In short, our plant juniper had already known the spotlight and been the focus of general human interest centuries before the Gin and Tonic—we really ought to be enjoying one right now as we read these lines, don't you agree?—came along. Yet, several theories exist regarding the origins of gin. For a long time, in fact, it was believed that the first person to create a juniper-based spirit was Dutch: a professor of medicine at the University of Leiden named Franz de la Böe, more famously known as Doctor Sylvius, in the 17th century. More recent studies have disproved this idea, however, highlighting the Mediterranean rather than Northern European origins of juniper spirits. Around the year 1100, the Italian city of Salerno boasted the world's most famous and innovative medical school. It could also rely on a fundamental instrument to produce any distillate, namely the alembic invented by "Gerber", a Persian alchemist, astronomer, and physicist whose real name was Jabir ibn Hayyan and who lived between 721 and 815. In the Salerno school of medicine's archives, several references to so-called "burning spirits" have been unearthed (also called "aqua ardens" or "burning water" in English). If we factor into this

THE MAN WHO DID NOT INVENT GIN

Although the invention of a juniper-based spirit, as has been established, is not owed to the Dutch scientist Franz de la Böe, he did in any case play a part in gin's story. Born in 1614 in Hanau, he completed a doctorate of medicine at the University of Basel and travelled extensively throughout Europe. He then became a professor of medicine at the University of Leiden, the oldest Dutch university, a position he held from 1658 until his death in 1672. In his numerous writings, he had the habit of signing with the pseudonym "Sylvius", from the Latin *silva* (meaning forest), which translated into Dutch becomes Böe. Among his many accolades, de la Böe was the first to note the role of tubercles in pulmonary tuberculosis. His name is given to the canal located in the midbrain (the sylvian aqueduct) and one of the main scissures of the cerebrum (the lateral sulcus or sylvian fissure). In many texts, he is mistaken for Franciscus Sylvius de la Boue, another noted doctor of medicine who around 1575 wrote about the "Eau de Genievre", meaning a sort of "water of life".

Though neither of these two can claim gin as their birthright, one can easily perceive how both men of medicine put forth the use of *aqua junipery* in their medical prescriptions, being men of medicine fully knowledgeable on the properties of alcohol and juniper. And while history does not tell us whether they added other botanicals to their distillates, nor does history deny it.

---

scenario juniper's well-established healing properties and the plant's abundance in the Campania region... well, to quote Agatha Christie: "A clue is a clue. Two clues are a coincidence. Three clues are proof".

In short, it's very likely that a juniper-based acquavite was being distilled in Salerno. It was, however, a kind of potion that the school's good doctors of the time would have intended as a curative, not an alcoholic beverage to enjoy among friends of an evening. And yet, not long after this period, the French count De Morret (son of Henry IV) was not only using juniper in "juniper wine" to flavor alcoholic drinks, he even acknowledged steeping juniper berries with other aromatic herbs in rye alcohol, to then distill the entire concoction. Back in Italy, a meticulous researcher named Fulvio Piccinino uncovered a text written by one Alessio Piemontese, the 1555 *Book of Secrets*, which contained a definitive, straightforward recipe: "And this is the liqueur" that truly contains juniper.

Though seen as a medicine or a means to flavor other alcoholic drinks, juniper was nevertheless on its way to achieving global fame.

The "Black Death" and "Dutch Courage"

From a medical standpoint, the success of gin's precursor was boosted by one of the greatest tragedies to ever strike the Old Continent: the spread of the bubonic plague, or "Black Death" as it was also called, which killed some twenty-five million people. The disease attacked the lymphatic system, particularly in the groin area. At that time, as previously noted, the diuretic properties of juniper were already known. Very quickly the use of the juniper berry grew significantly, along with the juniper plant itself, whose wood would be burned inside the house in the hope of dispersing the disease with smoke. Juniper's role in fighting off the plague is attested primarily by *aqua junipery*, a kind of "Crude Gin" in use sporadically until at least the late 1500s, then later more frequently; and even more so by a book written by Englishwoman Hannah Glasse in 1784, a century after a second wave of the epidemic struck London. In her book, *The Art of Cookery*, Glasse suggests a recipe for Plague Water, containing some twenty types of root, sixteen flowers, and nineteen different seeds. Not surprisingly, juniper is included among these.

But let us return to a more pleasant aspect of gin's story. Juniper-based spirits were certainly known to have other properties. In 1585, Sir Robert Dudley, 1st Earl of Leicester and a favorite of Elizabeth I (with whom he had a very close and affectionate relationship), was asked by the sovereign to support the Dutch in their battle against the Spanish. This was not merely a conflict of nations: it was a clash of religions that saw English Protestants taking sides with the Dutch against the fervently Catholic Spanish. While the military expedition was not a great success, Dudley returned to England with a nice tale to tell, one that would soon become legend. He told of the great courage shown by the Dutch infantry, which to him seemed to be boosted by heavy drinking just before battle. What they drank, in fact, was a juniper spirit, thus giving birth to the English term that would go down in history as "Dutch Courage". Dudley ended his career as head of the English ground forces and was then called to defend against the invading Spanish Armada in 1588. Luckily for him, the Armada never reached shore, but Dudley was not able to celebrate this, as he died shortly after from a probable stomach tumor.

It should be noted that in the era in which Dudley lived and encountered said "Dutch Courage", juniper was not entirely unknown in England. In fact, English distilleries used it as an aroma, for distillates to be sold alongside several other varieties, including the highly popular *aqua vitae*, aromatized with anise. Around 1570, such distillates were sold in "Strong Water Shops" and later "Dramshops", and by 1600 London boasted more than two hundred of these.

What Dudley had encountered with the Dutch, however, was not called gin, but rather "genever" or "jenever", which it is still called today. The next chapter will address the differences between the two products; here we are concerned with retracing juniper's journey towards becoming the spirit we know today. Genever's success in The Netherlands had begun in the 13th century, but it was during the so-called Dutch "Golden Age", from around 1600 to the end of the 1700s, that the spirit experienced its greatest fortune, helped greatly by The Dutch East India Company and its enormous fleet. In 1664 the very successful Bols company (still operating today), began to produce genever, which Dutch soldiers and sailors received in their daily rations. As logic would have it, "Dutch courage" soon made its way to England. Here it took root (coincidentally, it was often called "Hollands") for different reasons: the most immediate and simple of those being the percentage of Dutch people living in London at the time; the second was the semi-perpetual state of conflict between England and France that, at its most aggressive moments, brought trading blocks on all French products, including wine and spirits, highly popular in England; and the third was William III's ascension to the English throne in 1689, thanks to the parliamentary coup known historically as the "Bloodless (or Glorious) Revolution". William of Orange (whose full name was William Hendrick Van Oranje Nassau, but often referred to by the Scottish and Irish as "King Billy"), was in fact Dutch. He had, no doubt, favored the juniper spirit so popular in his native land. But as the English sovereign, he was determined to safeguard his country's interests as well as his crown's. His first act was to prohibit the importation of French brandy and to introduce economic bonuses for raw materials produced locally, persuading English distillers (who had supported his rise to power alongside landowners), to make their products using "good English wheat". His second was in 1690, with what would become known as the "Distillers Act", which permitted citizens to operate home distillation without requiring a license. All that was needed was a sign affixed outside one's house, ten days prior to opening for business. Additionally, around this time the king decided to raise taxes on alcoholic beverages, the most popular of which, up to that period in England, had been beer.

Gin as the "people's drug": the Gin Craze

The result was remarkable: from 1684 to 1710, while beer production dropped by some 12%, and stronger beers even more so, by 22,5%, production of juniper distillates grew by 400%. By 1684 consumption of *spirits* (gin, primarily) reached nearly two and a half million liters, a figure that grew to five and a half million or slightly more in 1700, to finally reach the astonishing amount of around thirteen and a half million liters. Some portions of the immense

BEARD, HAIR AND... DRAM

At the time in which drinking spirits was akin to a curative treatment, customers at barber shops such as the Worshipful Company of Barbers could receive a full package of treatments in one sitting: a haircut, a bloodletting, and a dram of alcohol, which almost invariably meant a glass of jenever. Such establishments, essentially health spas ahead of their time, would perhaps one day experience a revival—the bloodletting excluded, most likely.

---

quantities of gin consumed were even distributed as wages.

This was the beginning of the "Gin Craze". Several reasons contributed to gin's monumental success among the lower classes. In her 1977 article *The naturalization of Beer and Gin in early modern England*, Jessica Warner discusses two of these reasons: the competitive price compared to beer; and gin's flavor and alcohol content that induced feelings of euphoria that the average poor person couldn't possibly afford, typically (a poor diet of the time was based on stale vegetables, rotten meats, and bread whitened with alum, a mix of alluminium solfate, potassium, and gypsum).

With the per capita daily consumption soon reaching an astounding half-liter or thereabouts, one can easily see that things were getting out of control, even if the gin of that time had been top quality. But given that home distillation often relied on turpentine to flavor gin, alongside the competitive price and the general approval among the less well-off, the consequences of the Gin Craze were indeed terrible. The death rate in London rose enormously. For example, gin was sold everywhere during this period, from market stalls to pubs, and here the floors would be covered in straw to accommodate customers passed out from drinking. In places where gin was sold, one would often read the phrase: "Drunk for a penny, dead drunk for two pence, clean straw for nothing".

Reports from 1743 indicate that gin production reached around seventy million liters. Compare this to a population of some six million inhabitants. Keep in mind, too, that the low-grade gin of the lower classes was not that enjoyed by the well-to-do and noble classes. They, rather, continued to drink the Dutch version they called "Holland Gin". The poor, meanwhile, gave their versions of gin some rather colorful, folkloristic names, almost always speaking to its alcoholic effects: Madame Genever, Mother Ruin, Ladies Delight, Strip Me Naked, Slappy Bonita, Cuckold's Comfort.

The Gin Craze, immortalized by the famous 1751 William Hogarth print, lasted about forty years. And it was truly a craze: theatrical performances were often disrupted by the din of a drunken audience;

women would hide bottles of gin in their garments to then sell on the streets; much of the gin in question had on average volume of 80% (instead of the usual 40–45% volume of today); and crime and street brawls had become commonplace in the evermore populous city of London. Hogarth's "Gin Lane" (p. 58) offers a horrifying look at the pervasive squalor and desperation of the time. At the center a prostitute is pictured, a baby falling from her arms down the stairwell while a young boy fights a dog for a bone, a naked child cries as its mother is carried off on the undertaker's cart, and a cripple beats a blind man with his walking stick. By contrast, Hogarth's twin scene "Beer Street" depicts robust, hard-working people enjoying a pint of beer: the artist's clear condemnation of the London of his time.

The new Gin "habit"

In a word, corrective measures were needed. And they arrived between 1729 and 1751 with the English Parliament's issuing of some eight Gin Acts. Among these, the Act strongly pushed by Sir Joseph Jekyll and known as "The Fifty Pounds" is worth noting, so-called as producing gin now required an annual tax of said sum. Private distillers were subject to high fines and months of imprisonment. Such laws were respected thanks to severe enforcement measures. The Gin Act of 1736, a year in which London alone recorded some seven thousand shops selling lowest quality gin, or gin made with turpentine, sulfuric acid, alum or strychnine, was short-lived however, on account of numerous protests leading to its abolition in 1743. But the problem endured. Records from 1751 show a mortality rate exceeding the birth rate; gin was being unscrupulously consumed, even by pregnant women and thousands of children died of alcohol poisoning in London alone. The next Gin Act of the same year prohibited distilleries from selling to food shops and raised the beverage tax. Such measures, together with some years of poor grain harvests, forced many distillers to close, and slowly the river of gin running through the streets of the capitol began to diminish. In the meantime, serious professionals had entered into the gin business. In 1740, Finsbury was founded, followed two years later by Booth's and Greenall's in 1761. In 1769, a Scottish whisky distiller named Alexander Gordon opened his distillery in Bermondsey in South London, while in 1793 the Coates family lit their stills for the first time at the Black Friars Distillery in Plymouth, bringing the same-named gin to life. The market returned to a level of moral dignity and more moderate consumption, with a brief "flashback" in 1825 after the lowering of the *spirits* tax again. With the return of low-cost gin came the rise of the so-called "Gin Palace", buildings with large windows, long bars, and good lighting. While they didn't last long, Gin Palaces are credited by some with having influenced the interior style of later Victorian-era pubs. Then, when the Duke of Wellington

SAVED BY THE BELL

An interesting historical anecdote tells us that the expression "saved by the bell" dates to the same period as the Gin Craze. The fear of alcohol poisoning (coma) was in fact less feared than being mistaken for dead and buried alive. Thus the safety coffin was born, containing devices such as small bells that mistakenly buried people could ring once revived to alert cemetery staff. Instead of a bell, some of these coffins had observation windows that allowed for verifying the "conditions" of the presumed dead person. One of the oldest and most famous of these was designed by Duke Ferdinand of Brunswick, who died, beyond any doubt, in 1792. The coffin in question included a window for light, a tube for air passage, and a bolt that was operable from within.

Some safety coffins were even patented, and similar inventions continued to appear until the 19th century. Doctor Adolf Gutsmuth, inventor of one such type of coffin, trusted his invention so much that he allowed himself to be buried alive several times in order to demonstrate its efficacy. His colleague, meanwhile, one Johann Gottfried Taberger, designed a system of strings that were tied to the corpse and then to a bell above ground. Naturally, many false alarms rang, caused by the "movements" occurring as a corpse decomposed.

(champion at Waterloo in 1815 and subsequently Prime Minister of the nation), abolished the beer tax in 1830, and moreover permitted any citizen to open a retail shop by paying a mere two-guinea fee, the Gin Craze finally came to an end. Now entrusted to the hands of professionals, gin took on a new life. Select raw materials and botanicals laid the foundation for a top quality spirit, while advancements such as the column still invented by Aeneas Coffey in 1832 took care of the rest, guaranteeing a lighter, fresher product and paving the way for London Dry Gin. In short, English gin's first, crystalline fame came about in the 19th century, becoming part of the daily consumption habits of even the most well-to-do classes. It's telling, for example, that in the mid-century period, the Plymouth distillery was supplying the Royal Navy with one thousand barrels per year. Sailors were given daily rations of rum, and officers showed a strong preference for the more refined gin. The practice of issuing rum as part of British sailors' rations ended on 31 July, 1970, a day that history would come to call "Black Tot Day". The United States Navy had abandoned the practice much earlier, on 1 September, 1862; Canada, on the other hand, held out until 30 March, 1972. A "tot" was the daily ration of about 70 ml of 56.4% volume rum, and it was distributed via a precise ritual. At eleven in the morning, the boatswain's mate would whistle "Up Spirits", the call to the petty officer to head up the aft deck, to be joined by some of the Royal Marines officers bringing the keys to unlock the spirits room. The group would file in, checking that precisely the right dose of rum was taken from the cask, which was then administered to adult sailors and those not being held under punishment. Then the rum was taken up on deck. First the officers were served a ration of pure rum. Then the remaining rum was thinned with water and rationed to the sailors.

With a slight stretch of the imagination, we can credit the British Royal Navy officers with the invention of what would become history's first cocktail. For medicinal reasons, they were known to consume a few drops of Angostura Bitter (created in 1824), which had proven an excellent remedy for stomach cramps, digestive problems, and common colds, and they opted for mixing the bitters into a glass of gin, which immediately took on a delicate pink color. The Pink Gin was born. Another cocktail, the Gimlet, is similarly linked to the Navy. Long periods at sea meant poor nutrition, given the lack of fruits and vegetables available and thus needed vitamins, which in turn caused countless sailors to become ill and die of scurvy. In his logbook, British Commodore George Anson noted the devastating effects of the disease, which over the course of four years at sea virtually wiped out his crew. It was later discovered that citrus fruits like orange, lemon, and lime, were an excellent remedy for scurvy. But the problem was storing fruit for long periods on ships without any type of refrigeration. The solution came in the form of Rose's Lime Cordial, a lime juice with no added alcohol. When some officers of the Royal Navy thought about how to "enrich" this, the Gimlet was born.

Obviously, encouraging such quantities of gin wasn't done by just the British (or Dutch) Navy, who always made sure

A GLASS ON BOARD

Given its popularity on ships, gin became a "public relations" opportunity. Once a British ship was anchored at port, the crew would raise a white and green flag, an invitation to officers at port to come on board and enjoy a glass of gin, often Pink Gin. The flag took the name "Gin Pennant", "Gin Flag" or "Drinking Pennant". The practice seems to have become fashionable around 1940, and to this day is still practiced by any ship flying a Commonwealth flag.

enough genever was available and, indeed, always served it in large amounts—about a fifth of a bottle of overproof alcohol per person, per day. The 19th century saw the first bottles of gin being sold right next to the casks, the former quickly replacing the latter. And, to an even greater extent, this was the century that ushered in the first golden age of cocktails, which more than any other single factor would bring about gin's world-wide expansion.

The Cocktail Revolution

The origins of the word "cocktail" are frequently discussed, with several different opinions on the matter. The first appearance of the word on the printed page certainly seems to date to 16 March, 1798, in an article in London's Morning Post discussing a local pub manager who, having won the lottery, generously and somewhat astonishingly, offered to forgive all his clients' bar tabs. And the list of debts to be cancelled included a non-specified "cocktail".

A few years later, in 1803, the word popped up in the United States, and shortly after, in response to a reader's asking what the word meant, the editor of the newspaper Balance wrote: "A cocktail, then, is a stimulating liquor composed of spirits of any kind—sugar, water, and bitters".

The mystery surrounding its origins aside, the cocktail quickly occupied a growing space in lexicons as well as consumer habits. To what extent new drinks such as the Mint Julep, the Pegu Club, or the Pimm's Cup (created between 1840 and 1850 by one James Pimm in his London tavern and beloved by high society ladies, who in any case preferred to call it a "White Wine" or a "Nig"—gin spelled backwards—to avoid giving rise to malicious gossip) encouraged the growth of gin is difficult to say with precision. But that gin represented an ideal base for the heady expansion of cocktails to come is accepted by all. Confirmation of this lies in the fact that during the same century (the 19th) that saw the expansion of British gin, its older brother, Dutch jenever, also ushered in its golden age. The small city of Schiedam, for example, just outside Rotterdam, witnessed an increase in the amount of distilleries: from just 37 in 1700, to 250 a century later and some 392 by 1880. During those years, Bols jenever reached not only the rest of Europe but even Africa, Southeast Asia and, naturally, America. Even neighboring Belgium was taken up with the juniper trend, thanks to the concurrent imports ban on Dutch products during the war for independence and the column stills developed by Cellier-Blumenthal.

Yet, according to author Dave Broom, gin achieved its international fame thanks to the United States, initially taken with the Dutch jenever (which had crossed the ocean to land in America in 1750), then by Old Tom (the first truly English juniper-based spirit, which we will discuss further in the next chapter), and finally, but only at the beginning of the 20th century, Dry Gin.

WILLIAM HOGARTH: GIN LANE AND BEER STREET

The famous prints titled "Gin Lane" and "Beer Street" (p. 58), among the most compelling admonishments in the battle against the British Gin Craze, were created by William Hogarth, an artist who lived in London between 1697 and 1764. Painter, engraver, and printmaker of satirical works, Hogarth read the news regularly and clearly discerned the reality of his time, creating drawings that struck the hearts and minds of readers. "Gin Lane" and "Beer Street" were published in the London Evening Post between 14 and 16 February, 1751, representing for Hogarth something more than a mere artistic endeavor. He, like so many Londoners, had been affected by the Gin Craze: it had taken the life of one of his dearest friends.

The Gin Lane

The work is set in the parish of St. Giles, a notorious slum that Hogarth depicts in several of his works. It depicts the squalor and desperation of a community that seems to revolve exclusively around gin. Spiritual poverty, sordidness, death, and decay pervade the scene. The only flourishing businesses are those that serve the gin industry: sellers, a distiller named Kilman, the miser Mr. Gripe who greedily strips the street's alcoholic residents of their possessions in exchange for a few pennies to feed their habit (a carpenter is pictured selling his saw, while a housewife offers up her cooking utensils), and the undertaker, whom Hogarth has provided with a few new clients in his scene.

The central focus of the image is a woman entirely overcome by gin consumption, who has been driven to prostitution (as evidenced by the syphilitic sores on her legs), and who lets her baby slip out of her arms and fall down the stairwell to the gin cellar below, to what will surely be death. Half naked, the woman's only concern is her next pinch of snuff. The depiction of this mother was not such an exaggeration as one might think: in 1734, a woman named Judith Dufour strangled her two-year-old baby, removed the baby's clothes and left the body in a ditch to then be able to sell the clothes to buy gin. In another case, an elderly woman named Mary Estwick let a baby burn to death while she slept in a gin-induced stupor. Such episodes fuelled the flames of the anti-gin movement and its activists, such as the tireless Thomas Wilson, as the image of the negligent mother grew evermore central to the anti-gin propaganda.

Other images of despair and madness fill this scene: a drunken man on horseback beats himself with a set of bellows, while holding a baby impaled on a spike; the mother of the baby rushes from the house, screaming in horror; a barber has taken his own life in the rundown attic of his shop, his business in ruin because no one can afford to get a haircut or shave; on steps below the woman who has let her baby fall, a skeletal pamphlet vendor (perhaps a former soldier who sold his close to buy gin and has died of starvation), is holding a basket of moralizing pamphlets titled *The Downfall of Mrs. Gin* on the evils of gin-drinking; next to him sits a black dog, symbol of depression and despair, while outside the distillery a fight has erupted and a mad cripple lifts his crutch to strike a blind compatriot.

Even children seem to be victims of the Gin Craze: one is being calmed by its mother with a cup of gin and in the background, an orphan baby wails, naked on the ground, while the body of its mother is loaded into a coffin. Two girls from the parish of St Giles, as indicated by the badges on their arms, each drink a glass of gin. In front of a door, a starving boy fights a dog for a bone, while next to them a girl has passed out. Approaching her is a snail, symbol of the sin of sloth.

In the distance, the church of St. George's in Bloomsbury can be seen, while above, the pawnbroker's sign forms the shape of an upside-down cross perpendicular to the church tower, signifying the corrupted worship of the gin drinkers.

Beer Street

Here all is joyous and prosperous: industry and happiness go hand in hand.

The only business in trouble is Mr. Pinch's pawnbroker, with Mr. Pinch himself living in a crumbling, rundown building. In contrast with Gin Lane, the prosperous Gripe displays expensive glasses in the upper window (a sign of his flourishing business). Pinch, on the other hand, displays only a wooden contraption, perhaps a mouse trap, in his upper windows, while he is obliged to take his mug of beer through a window in the door, suggesting that his business has been so unprofitable as to put him in fear of debtor's prison.

The rest of the scene is full of English workers bustling about in good humor. It is 30 October, the birthday of George II (indicated by the flag that waves in the background atop St Martin-in-the-Fields church), and the locals in the scene are surely drinking to the king's health. Under the auspicious Barley Mow sign, a blacksmith holds a tankard of foamy beer with one hand and a leg of beef with the other. Together with the butcher, he laughs with a cart driver as he courts a waitress (the key in her hand as a symbol of domesticity). Ronald Paulson suggests a parallel between the trinity of ill-omen signs presented in Gin Lane (the pawnbroker, the distiller, the undertaker) and the trinity of English men in Beer Street (the blacksmith, the cart driver, the butcher). Nearby, a pair of fishmongers rest with pints in hand and a porter sets down his load to take a break. In the background, two men carrying a sedan stop for a drink, while the passenger remains wedged inside: her large hoop skirt holding her fast in place. On the roof, builders working on the pub owner's house above the Sun Tavern share a toast with the owner of a tailor shop. In this image, a barrel of beer hangs from a rope above the street, in contrast to the body of the barber in "Gin Lane".

The inhabitants of Beer Street and Gin Lane are drinking rather than working, but the former are resting after their labors (all those depicted are at their places of work, or have their goods and tools of their various trades nearby), while the latter are clearly drinking immoderately instead of working. Hogarth's condemnation is evident in the raw yet realistic scene depicted in "Gin Lane". Not coincidentally, these images have been reissued innumerable times, providing us with an example of how the excessive use of alcohol not only harms the individual but also demeans all of society.

The Americans even began to produce gin, Broom suggests. And for one simple reason: they loved to drink cocktails. It's interesting to observe the changes in taste with regard to gin through celebrated, historical cocktail books. The first, a legendary collection of recipes written by Jerry Thomas in 1862, saw gin as jenever or Old Tom. A prediction that was confirmed in the subsequent 1888 edition as well as in his fellow barman Harry Johnson's book. It was only in 1908 with the publication of *The World's Drinks and How to Mix Them* by William "Cocktail" Boothby that the first Dry Gins appeared. The 20th century seemed primed to be just as successful for gin as the century prior.

THE EIGHT GIN ACTS

Between 1729 and 1751, the British Parliament issued some eight Gin Acts, to tax spirits and establish license costs and rewards for whistleblowers.
Here are the eight Gin Acts and the year of their enactment:

The first Gin Act – 1729
The second Gin Act – 1733
The third Gin Act – 1736
The fourth Gin Act – 1737
The fifth Gin Act – 1738
The sixth Gin Act – 1743
The seventh Gin Act – 1747
The eighth Gin Act – 1751

The most famous of these was the Gin Act of 1736, also known as the "Fifty Pounds". It was passed in parliament thanks to Sir Joseph Jekyll (who was utterly horrified by drunkenness) and established a tax equivalent to twenty shillings per gallon, while gin producers were required to pay an annual tax of fifty pounds to obtain a license. In the following seven years, only two such licenses were registered. Furthermore, the act in question provided for a hefty fine and two months imprisonment for illegal distillers, a financial incentive for those who informed authorities about "unlicensed" stills and, above all, the police forces needed to enforce these laws. Hence, the cost of a glass of gin rose from a penny to the equivalent of a week's rent. A collective revolt ensued: people took to the town squares and gin shops hung black fabrics on their signs as a form of mourning. Throughout England, organized funeral processions took place to honor the death of "Madam Geneva", and Henry Fielding, author of the novel *Tom Jones*, wrote a protest pamphlet. The act proved to be a complete failure and was abolished in 1743.

The government found itself forced to take further action, namely a new Gin Act that raised beverage taxes and prohibited the sale of distillates to food shops, prisons, or factories. Consumption dropped drastically for the rest of the 18th century.

Despite the high licensing fees and taxes imposed by the Gin Act of 1736, by 1750 some twenty-nine thousand authorized gin sellers were registered throughout England—and one can only guess as to how many unlicensed shops were in operation at the time. By social standards, drinking gin was still safer than drinking water, and thus it remained rather normal to see children drinking it alongside adults.

From Prohibition to the "golden age"

On 16 January, 1920, despite President Wilson's veto, the United States Congress passed the so-called "Volstead Act". It was the beginning of thirteen disastrous years, fueled by various "temperance" movements (one of the more famous being the Blue Ribbon Party), often driven by women and religious groups, years that history would come to associate with one word: "Prohibition". By making the sale of all forms of alcohol illegal, America set out to end alcoholism and all its related consequences. The results were quite varied: not only did the number of alcoholics grow, but the alcohol market also ended up in the hands of organized crime, while the network of secret underground establishments, the famous speakeasies, quickly surpassed the legal bars in operation before 1920. Even those who didn't frequent these clandestine bars drank. Among the more unusual things Prohibition brought about was homemade gin, also known as "Bath Pipe Gin" as it was often made in home bathtubs, and was easier and cheaper to make than other spirits. This gin was never of a fine quality though, and in many cases it meant, more than anything else, a shortcut to the morgue. Even more fascinating, after the abolition of Prohibition in 1933, gin emerged without having suffered any damage to its reputation. According to some, in fact, it was actually during Prohibition that gin became a truly popular spirit. Of course, it was not on account of gin that Prohibition was abolished but instead due to the absolute pointlessness of a law that had created more problems than it resolved. Indeed, in some ways those thirteen years of false sobriety were America's own Gin Craze.

The only difference is that between 1920 and 1933, American tastes changed, shifting from sweeter, more delicate gins such as the Dutch jenever and Old Tom to drier gins. The 1940s and '50s then brought the final confirmation of Dry Gin's reign and its most iconic cocktail: the Dry Martini. At this point, nothing could hinder the global success of this juniper-based distillate that had been born as a curative for various ailments. A versatile component of countless cocktails, beloved by famous barmen and celebrated writers like Ernest Hemingway, singers such as Frank Sinatra, and politicians like Winston Churchill, gin could, by now, virtually rest on its laurels.

Gin could not have known, however, that it would soon meet its match, and in its own "field". This time, the setback was not on account of taxes or political priorities, but for the sudden, exuberant appearance of vodka, which in the mid-1950s conquered the market with its two key advantages: its likewise great adaptability in cocktail making, and its supposedly less invasive odor. In other words, compared to gin, vodka left less of a trace and was less offending, once consumed. In England, one of gin's birthplaces, drinking gin suddenly meant belonging to an older generation, those who frequented golf resorts and sailing clubs. And certainly the youths of

Swingin' London couldn't drink what their parents were drinking, just as they couldn't wear the same clothes or listen to the same music. So while vodka didn't exactly kill gin, it definitely put gin in its place, leaving it bereft of its long-enjoyed allure. It remained useful, but no longer indispensable. And it lost, if you will, its prior glamour.

But in 1987, something strange happened. One Sidney Frank introduced a new gin to the market, labeling it "premium". He packaged it in elegant, blue glass bottles that he filled with a distinctly aromatic spirit, light and highly stimulating to the senses (including the visual). He called it "Bombay Sapphire", and just like that, gin's flame was relit.

THE DAZZLING GIN PALACES

Around 1820, thanks to a lowering of the spirits tax, England witnessed a gin revival. It began with the so-called "Gin Palace" era, short-lived though utterly dazzling. In a London lit by gas lamps (which replaced oil lamps in 1807) and increasingly populated (with nearly two million inhabitants by 1841), Gin Palaces were regarded for their large windows, long mahogany bars (often gorgeously engraved), and lighting. They even caught the attention of Charles Dickens, who described them as "perfectly dazzling when contrasted with the darkness and dirt we have just left" and labelled them "seduction personified". Taking the place of the more dreary Dram Shops, Gin Shops functioned more like a take-away than a gathering place. The first Gin Palace was the beautiful Thompson & Fearon's on Old Street, built in late 1829 and designed by architect John Buonarroti Papworth. The Gin Palace was significant also for the more visible roles their bartenders played, and their more "modern" relationship with customers. Additionally, Gin Palaces set the style for later establishments, the Victorian pub that we all now know. Today several examples can be admired in London and elsewhere: some examples from the late 1800s include The Princess Louise in Holborn, The Princess Victoria on Uxbridge Road, the King's Head on Upper Tooting Road, the Baker's Vault in Stockport (which has kept the vaults under which gin casks were once housed) and the Crown Liquor Saloon on Great Victoria Street in Belfast, now restored and run by The National Trust.

THE FIRST COCKTAILS: PINK GIN, GIN PAHIT AND THE GIMLET

Calling them "cocktails" today might seem a bit extreme. Yet these drinks were certainly the precursors to the custom of mixing different ingredients in a glass, laying the ground for mixology. Pink Gin, for example, was born as a means to ingest various bitters considered medicinal aids for digestive problems and the prevention of some diseases. Angostura, Rack's, and Peychaud's all shared a bitter flavor, and adding these to what was deemed the purest spirit around, namely gin, was only natural. When added to gin, the dark and murky bitter turned it a much more pleasing pink color, hence the name "Pink Gin". Typically served at room temperature, this drink also had a non-alcoholic version known as The Campbell, which called for lemonade instead of gin. Reverend Stoughton, father of the renowned Stoughton Elixir, recommended the astounding dosage of 50/60 drops of his product in a glass of water, beer, or spirits.

Thus the passage from softening bitter flavors in gin to flavoring gin itself was a short one. Meanwhile, Indian Civil Service officials used to add pickled onions or chili to their glass of gin, renaming it "Gin Pjai". Malesia, on the other hand, gave us the Gin Pahit (in the local language pahit means "bitter"), with a bitter-to-gin proportion of one-to-three. Beloved and alluded to by writers such as Somerset Maugham and Graham Greene, the Gin Pahit is today known also as the "Dry Bone Churchill Martini".

Being at sea for many months brought with it the great problem of scurvy, a disease caused by a lack of vitamin C and one that could wipe out entire crews. Once the vitamin's presence was discovered in citrus fruits like lemons, oranges and limes, administering these fruits became the key remedy against the disease. This, however, was not so easy in practice, as the surgeon James Lind would soon discover: in 1747, after having confirmed the properties of citrus fruits and challenging prior beliefs that recommended vinegar or a low-salt diet, Lind found himself facing the issue of long-term storage of these fruits. His first attempts involved boiling the citrus, yet this did not eliminate the risk of fermentation, while another early solution was to add 15% of Demerara rum to the juice. In 1795, the admiralty suggested Lind's solution, even though the amount of rum needed forced officials to review the daily spirits rations given to sailors. Consequently, a mix of rum and lime became very popular, so popular that soon British sailors would come to call it a "Limey".

In 1852, a kind of medicine called "Gin Mixture", similar to the modern Gin Sour, was beginning to be served on board ships of the British Royal Navy. Some years later in 1867, the Ministry of the British Merchant Navy published a note officially introducing lemon and lime juice as a remedy to fight scurvy on board ships. But the real turning point came around 1865, when a British merchant named Lauchlan Rose patented his Rose's Lime Cordial, a non-alcoholic product that contained high amounts of lime and could be kept for long periods. The lucky Lime Cordial not only resolved the problem of scurvy on board ships, but also because it gave life to the Gimlet, was one of the first cocktails to appear officially in a text from that time period. Its first reference dates to 1917, in the *The Ideal Bartender* by Tom Bullock, who called it the "Gillette Cocktail", comprised of equal parts lemon juice and Old Tom, plus half a teaspoon of sugar. The Gimlet then earned its place in *Barflies and Cocktails* by Harry MacElhone in 1927 (equal parts Plymouth Gin and Rose's Lime Cordial) and in the famous *The Savoy Cocktail Book* by Harry Craddock, published in 1930.

According to legend, the Gimlet owes its name to the surgeon Thomas Gimlette, who served in the Royal Navy from 1879 to 1913. Other accounts suggest the name derives from the corkscrew-like tool used to open lime juice containers, also called a gimlet.

In the following years, it aided the resurrection of Plymouth Gin, fed the legendary status of master distiller Desmond Payne (the man behind Beefeater Gin), and played a part in the birth of a truly creative frenzy, one that saw thousands of small and medium distilleries springing up like mushrooms on every continent, in every nation, and in almost every city. This ongoing phenomenon accounts for the subsequent "dusting off" of historic gins like Old Tom and jenever, and is rooted firmly in the characteristics of gin: those commonly known, those discovered or rediscovered only recently—yet all intrinsically part of this remarkable spirit's genetic heritage.

Gin unites syrups, bitters, and fruit juices, while also expressing the qualities of terroir (consider, for instance, how many gins today use local herbs, spices, and roots), suggesting that the journey of this small, robust, aromatic berry will not be ending any time soon. Perhaps it will encounter more dips and turns in the road, or the occasional challenging climb, but gin will no doubt continue onwards, towards an elusive horizon.

Chapter II

The world is gin

"If I had a thousand sons, the first human principle I would teach them should be, to forswear thin potations"

William Shakespeare

A significant part of gin's fascination for humankind is due to its long and troubled history, and even those who know very little about the subject are likely to recall some gin-related anecdote or incident from that history. Another part is certainly due to the generations of barmen who have created and who continue to create an infinite amount of gin-based cocktails and long drinks. A third part to consider also exists, and that is the diversity one finds embodied within a bottle of gin, in large part thanks to an explosion of micro-distilleries and the creativity of the master distillers who have been liberating themselves to an unprecedented degree. Yet the current expansion of the world of gin, evidenced on the bottle shelves of any cocktail bar, rests, in a manner of speaking, in historical diversity. Because when it comes to gin, or rather juniper-based spirits, we should always speak in the plural, if for no other reason than to avoid doing wrong by the five gin cornerstones we describe here, in strict order of their arrival in the world.

JENEVER
The "father" of gin

Gin's oldest ancestor is Dutch. Yet despite the clear connection between jenever (or genever) and The Netherlands today, this early famous juniper-based spirit actually evolved in the Low Countries, an area which comprised Holland, Belgium, Luxembourg, with the first news regarding gin originating in the Belgian city of Bruges. In the 13th century, Bruges was a lively city and important crossroads for traded goods, among which was "gruit", a mixture of herbs and spices used for centuries to flavor beer, before the global popularity of hops took hold. Naturally, this Bruges drink was not yet the classic jenever; it was considered more a type of brandy (a modern contraction of the word "brandewijn" meaning "burnt wine"). During this time, the word "brandewijn" referred to any type of fermented alcohol, whether derived from fruit or from grains, which was then "burned"—in other words, distilled. Juniper's role was not yet predominant; it would become so later, when the plague epidemic swept through Europe. In 1349, the plague struck the Low Countries and, in that atmosphere of widespread panic, juniper seemed an excellent medicine. Doctors of the time would move through the sick covered entirely by wax suits. On their faces they wore masks shaped like birds' beaks, equipped with openings and small screens for the eyes, and two small holes for breathing. The tip of the beak was filled with scented herbs and spices, including juniper but also lavender, thyme, mint leaves, and often a vinegar-soaked sponge. The goal was to create a protective filter against the miasma present in the air.

Juniper wood and bark were burned in home fireplaces in order to diffuse an aromatic and protective smoke throughout the rooms. With juniper ever-present, an increase in its use in the distillation of spirits was only logical as a consequence. And evidently the flavor was also very pleasing, which meant that after the health emergency subsided, the spirit remained.

In 1552, a certain Philippus Hermanni of the Constelijck distillery mentioned an acquavite "genever" to indicate a juniper-infused brandy. Some years later came more talk of an "Aqua Junipery", and for the first time the reference was specific to a grain spirit. Then, in 1575, the Bols distillery *t Lootsje* opened near the outskirts of Amsterdam and began producing brandewijn, by a family who apparently had learned the art of distillation in Cologne. It was only towards 1664 that they started to focus on barley spirits aromatized with juniper, a clear sign that the market was starting to favor jenever. Meanwhile, the start of the 17th century saw the rise of the Dutch East India Company, a mighty naval fleet of some five thousand vessels. This was a golden century for Holland, during which it would become one of the richest nations on earth and attain a near-absolute monopoly on international trade. In my view, the

Company was a notable driving force for jenever, destined to become the first juniper spirit exported and consumed throughout the world. Sales spurred production, obviously, and in the course of a few years, jenever distilleries virtually exploded. In 1658, the Onder De Boompjes opened its doors, followed by the small Van Wees distillery in 1679, which until 1970 was still supplying bottles of jenever to bars and restaurants. Then, in 1691 it was Nolet's turn, better known as Ketel One, opening in the Rotterdam suburb Schiedam, an area with more than four hundred active distilleries, known throughout the 19th century as the "capital of jenever" (incidentally, the Jenever Museum is worth a visit; located at Lange Haven 74–76, Schiedam). The list goes on: Wenneker (1693), de Kuyper & Zoon (1695), Rutte & Zoon (1830), and Van Toor (1883).

The Second World War, later competition with other spirits (including Dry Gin), and declining consumer preference all contributed to a reduction in distilleries. But jenever did not disappear. In 2008, it received protected status from the European Union as a typical product of Holland, Belgium, and some French and German provinces. But what is jenever, exactly? Firstly, jenever belongs to the category known as compound spirits, which blend two different products: the so-called "moutwijn" or "maltwine" and a neutral spirit flavored with herbs. Moutwijn is typically produced by just one distillery, Filliers Distillery, starting with a neutral grain spirit in most cases, using double distillation, and not column stills. Moreover, differently from gin, jenever is obtained from a blend of grains, such as corn, rye, and barley. Percentages vary depending on the individual distillery's indications, and according to the production philosophy and the final desired flavor. The flours are poured into the boiling pot in a very specific order: first the corn when the water reaches about 95 °C, then the rye, making sure to lower the temperature by fifteen or twenty degrees, and finally, when the mixture is at 65 °C, the barley, a portion of which doesn't have to be malted. After cooking for two or three hours, the mixture is left to cool to 20 °C. Then the yeast is added. Once the fermentation is finished, the mixture is transferred to the still. It should never be filtered, which is about more than simply saving time; it's fundamental to preserving bread and malt hints, characteristic qualities of the finished product.

Next comes distillation. This can be done in two different ways: for the highest quality jenevers, medium capacity pot stills are used, and the distillation happens two to four times. More commercial types of jenever use short column stills, those with about twelve to sixteen plates, and are double distilled.

The herbal distillate, which naturally contains juniper, is closer to how London Dry Gin is made, being essentially a neutral grain spirit similar to a vodka, re-distilled with juniper and other botanicals. It's interesting to note that Dutch legislation

categorizes "jenever" as a distillate that must contain juniper, but, in contrast to British legislation (which stipulates a predominance of this botanical over others), the Dutch are content with the mere presence of it. Even if this means a single juniper berry.

Finally, keep in mind that all the different categories of jenever, including this one, should be considered in terms of plurality. Take, for instance, l'Oude Jenever (whose name doesn't mean old in this case, only traditional), a type of jenever that must contain a minimum of 15% moutwijn and be at least 35% volume, with a maximum of 20 grams of sugar per liter. It can be aged if desired, and if so will clearly indicate so on the label. The aging must be at least one year, in barrels of maximum seven hundred liters. Then, if the word "graangenever" or "grain genever" is found on a label, this means the moutwijn has been obtained exclusively from corn.

Jonge Jenever is a jenever rather on the young side (in a manner of speaking), having come out in 1950. Its flavor is lighter than the flavor of Oude, associated more with a malty component than juniper. To be categorized as Jonge, the percentage of moutwijn mustn't exceed 15% (more often it doesn't exceed even 5%), the minimum volume must be at least 35%, and the amount of sugar equal to maximum 10 grams per liter. Here, too, the word "graangenever" means obtained solely from corn. The third type of jenever to get to know is korenwijn, which has a slightly higher alcohol content (minimum 38% volume) and a sugar content never more than 20 grams per liter, but the percentage of moutwijn is decidedly greater, the minimum of which is actually fixed at 51%. Korenwijn and Oude Jenever are both suited to aging, with the first improving greatly over time and the second, the Oude, achieving its best result with a maximum five years of aging.

Finally, we have the Fruit Jenevers. These differ greatly in flavor from the classic versions on account of the fruity notes that tend to become the characterizing note. Remember, too, as we conclude these pages dedicated to jenever, that aging must take place in casks that previously held bourbon or cognac. The production year can be applied as long as the maître de chais does not add any flavored alcoholates at the end of the aging process. The terracotta bottles in which jenever is often contained are a nod to tradition, this economic and durable material having once been preferred to glass. Some distilleries let the spirit rest for some weeks (and up to three or four months) in large terracotta or ceramic amphorae before bottling.

The classic cocktails: Gin Julep, Holland Gin Cocktail, Improved Holland Gin Cocktail, John Collins

ORIGAMI

TYPE	ALCOHOL CONTENT	TECHNIQUE	GARNISH
OLD STYLE Old fashioned	31% abv	Stir&Strain	With lemon peel or a small origami

RECIPE (8 cl)
2 cl Genever NOTARIS Bartender's Choice
3.5 cl Puni Bianco Whisky
1.5 cl paper syrup
A few drops of HELLFIRE BITTER
Drops of chocolate bitter
Drops of Maraschino Lazzaroni

PREPARATION
Paper syrup
Making paper syrup isn't easy, given that it contains non-edible glues. I therefore opted for a preparation that recalls the flavor of paper, consisting of rice starch, maraschino, Chartreuse Verde, and gum arabic syrup.

METHOD
Note that this drink is perfect in the Old Fashioned style, but here I prefer to make it directly in the mixing glass (either glass or metal), to be able to control dilution resulting from temperature changes, as it must absolutely remain very cold. Eliminate the excess water. Pour in all the ingredients, being careful to keep the smaller proportioned ingredients for last, as those work as adjusters and as a result need more attention.
Serve in a small goblet cocktail glass, as the Origami's perfect serve calls for a large chunk or ball of ice in order to keep the drink as cold as possible, while maintaining a consistent, light dilution.

OLD TOM

The mystery of the cat

Gin lovers are known to ask, Who was Old Tom? While the history of jenever may be clear, Old Tom's still poses many doubts, especially given that defining it as a "style" (note my intentional quotes) seems a stretch. What we can state with certainty is that the mysterious Old Tom represents the link between Dutch jenever, the version similar to that produced also in London, and, finally, London Dry.

Many tales speak to the origins and meaning of the term "Old Tom", an old-fashioned way of labeling any male cat. One of the more intriguing, though less realistic, tells the story of a certain captain, Dudley Bradstreet, a fictional folk character who might have been born in 1711. A textiles merchant and brewer, he was also an army soldier and undercover government official who passed information gathered on clandestine distilleries to authorities. At the same time, Bradstreet himself was an illegal producer. Apparently Bradstreet had rented a somewhat hidden-away place on Blue Anchor Alley, outside of which he hung an amusing sign depicting a black cat. But the cat sign actually hid an ingenious contraption: by dropping a few coins into a front slit, one could then push a lever and receive a dose of gin in exchange. This type of automatic gin distributor quickly proved very popular, and the cat sign, our future "Tom", entered into local slang as a word for gin.

It was a very ordinary gin, mind, nothing like the sweeter gin enjoyed today. In fact, several writings uncovered by researcher and writer Gary Regan show that throughout the 1700s and up until the start of the 1800s, "Old Tom" referred to any generic gin, without any specifics on its sweetness or lack thereof. And there's no actual proof that the gin Bradstreet's customers drank was sweet. Additional support for Regan's thesis is found in a 1910 edition of the Encyclopedia Britannica, which reports on a 1903 judicial hearing (Boord & Son v. Huddart) addressing ownership rights of the Old Tom trademark. During this hearing, in an attempt to establish a claim, one of the two plaintiffs purported that a cat had accidentally fallen into one of the gin vats. As astonishing and barely credible a tale as Bradstreet's, yet one that undoubtedly renders the history of gin more "colorful". More tales exist, but the important thing to note about this legal case is that neither party characterizes Old Tom as a sweet gin.

Most likely, the shift to a sweet version of gin was gradual. In a pamphlet from 1901, wonderfully titled "Mamma's Recipes for Keeping Papa Home", an advertisement for David McArthur & Co's celebrated cordial Old Tom Gin describes it as having a "very fragrant and agreeable odor and taste, characteristic of Gin of good quality...The spirit is of considerable strength, and has been somewhat sweetened". Hugh Williams, famous master distiller for producers of such caliber as Booth's, Gordon's, and Tanquer-

ay, affirms that a sweet Old Tom gin was produced beginning only at the start of the 1800s, alongside unsweetened versions. But why sweeten gin? We have some hypotheses to put forward: one of the more believable is that, at least in the early years, sweetening gin helped to hide defects in distillation. A confirmation of this comes from an illegal seller from the time, who called it the "poisonous dram". During the Gin Craze, a large part of the gin in circulation contained very little true gin. Many clandestine distillers had no qualms about using turpentine (today used to thin paint) or sulfuric acid. Sweetening this alcoholic liquid could have been an effective way to make it more drinkable.

Obviously the Old Tom produced from the start of the 19th century by serious, professional distillers did not incur any risks. Sweetening gin was achieved in different ways. The most common was to add sugar

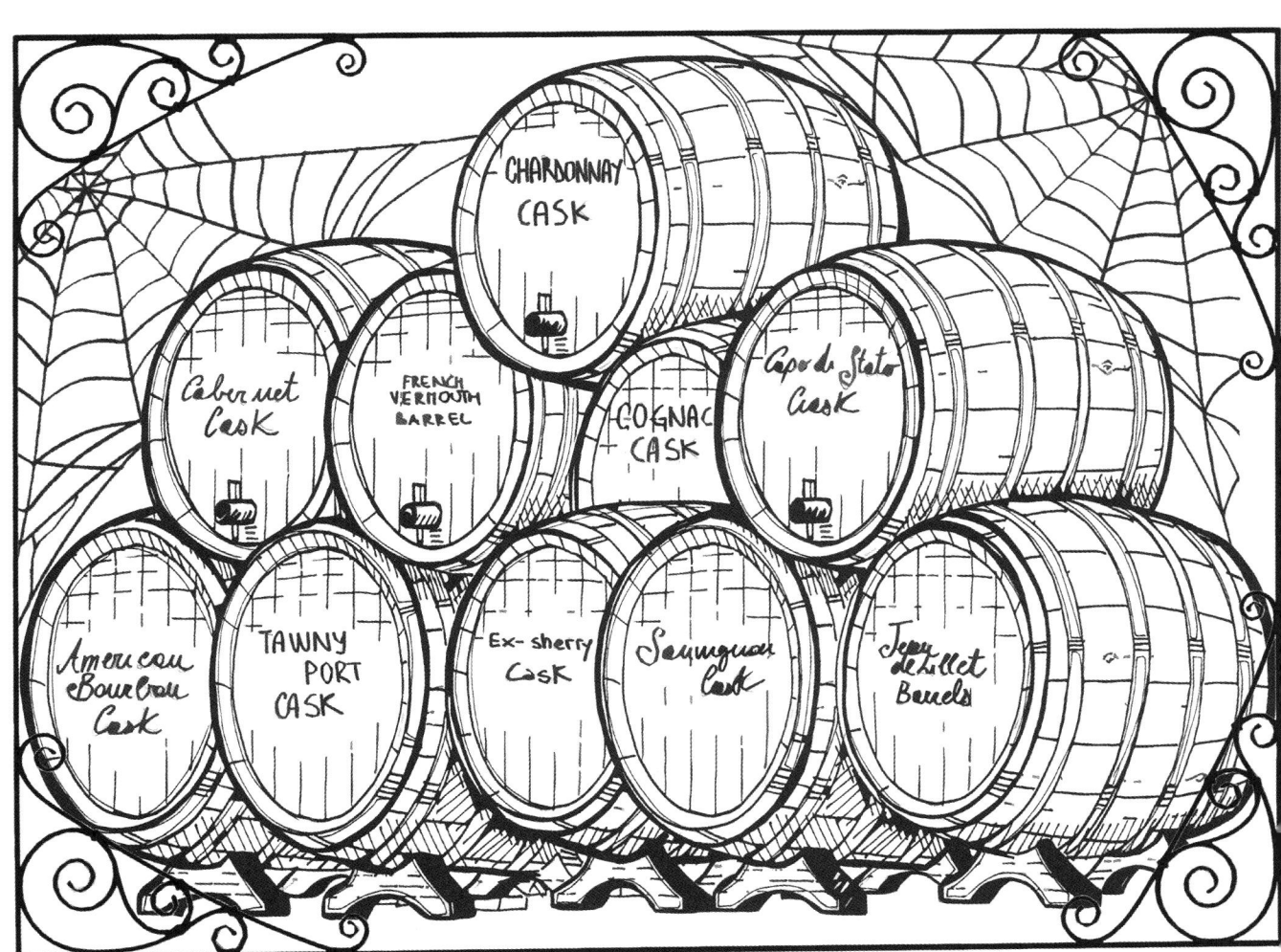

to classic gin ("sugar-sweetened"), such as in the case of Gordon's Old Tom. Then, another method did not include sugar but instead added a sweet final note thanks to the choice of select botanicals ("botanically sweetened") like vanilla, liquorice, and ginger. For example, the Old Tom made by Jensen's and Secret Treasures, and the Tanqueray Old Tom. Many American craft distillers add sugar to their Ould Tom (spelled in this manner), but this is with a base more similar to jenever than classic gin. And often the spirit is left to rest in barrels for a certain period of time. Some examples include Corsair's Major Tom, Spring 44, Downslope, Sound Spirit, and Ransom. Finally, it's worth noting that some gins labelled Old Tom are only such in brand name, not in style. These include gins such as Wray & Nephew's or Boord's.

The final question that can now be addressed, and one that to me is more useful than knowing the precise origin story of Old Tom, is whether or not it is possible to "self-produce" an Old Tom. By simply adding sugar? Yes and no. Simply adding sugar would not suffice—sugar being insoluble in alcohol—and thus an intermediate step is needed: namely, dissolving the sugar in water in the desired portions and adding it to gin, knowing that doing so will lower the alcohol content. It's a good idea to dust off some mathematical knowledge here, and perhaps equip oneself with a calculator while making your first attempts. Lastly, it goes without saying, but it can't hurt to specify: having a cat nearby is not a guarantee of success.

The classic cocktails: Original Martinez, Old Tom Gin Cocktail, Tom Collins, New Orleans Fizz alias Ramos Gin Fizz

TOP HAT

TYPE	ALCOHOL CONTENT	TECHNIQUE	GARNISH
OLD STYLE Boulevardier	29% abv	Stir&Strain	To your liking, with grapefruit peel/quality cherry

RECIPE (8 cl)
3.5 cl OLD TOM GIN
1 cl Bitter Campari CASK TALES
2 bsp Varnelli Mandarin Liqueur
2.5 cl Cinzano 1757 Vermouth di Torino (Red)
1 bsp Chartreuse Yellow

NOTES
A bar spoon (bsp) is equivalent to $1/8$ oz, or 0.37 cl.

METHOD
This drink can be made using a mixing glass (glass is fine, but metal is common) to control dilution owing to temperature maintenance. Everything should be very cold.
Eliminate the excess water. Pour in all the ingredients, being careful to keep the smaller proportioned ingredients for last, as those function as adjusters and as a result need more attention.
Pour the contents into an elegant, very chilled cocktail glass, or serve it in a Samadoyo Mug, adding a small piece of grapefruit or lemon peel, one star anise and a raspberry.

PLYMOUTH
A distillery, a style

Plymouth, in County Devon, England, is located at the mouth of the rivers Plym and Tamar. It was the departure point of The Mayflower, aboard which the Pilgrims travelled to America, was bombed in 1941 during the Second World War, and is also home to the largest Royal Navy base. Today it is still an important commercial port, but, for our purposes, Plymouth is first and foremost the birthplace of a style of gin that takes its name from the city and seat of a unique distillery. The establishment in question was founded by the Coates family in 1793, within a former Dominican monastery dating to 1431 (yet there are those who challenge this detail, such as the author Tristan Stephenson). Some records housed by the company indicate that distilling has been practiced there since 1690, a date that makes Blackfriars Distillery (or Coates), the oldest in the United Kingdom. The company changed hands for the first time in 1975, when it was acquired by Whitbread. Then, in 1996, it was bought by the Allied Domecq Group, which, in 2005, sold it to the Swedish V&S Group, makers of Absolut Vodka, among other products. Finally, in 2008, the distillery landed in the hands of Pernod Ricard, who had purchased the V&S Group.
These changes of ownership aside, the history of Plymouth Gin is truly fascinating. Plymouth Gin is more than a mere brand—it's a true and proper style. And it is recognized as

THE MAYFLOWER

The Mayflower, the ship destined to become an icon of American history, set sail from a River Thames mooring in July of 1620. Aboard were some one hundred Pilgrims (the exact number continues to be a subject of debate among historians), whose destination was the Virginia coast but who landed instead in what is currently New England, on 11 November, 1620. They settled in a colony that would come to be known as Plymouth, in honor of the last port they'd seen before crossing the ocean. During their arduous first winter, they established good relations with local native tribes, relations that would be celebrated the following year with a communal feast. The day has since become one of the most important holidays on the American calendar, Thanksgiving.

such by virtue of its Protected Geographical Indication status, one of the few in Great Britain, at least until 2015 when Pernod Ricard decided that it was no longer in their interest to maintain the property. It can be made only in Plymouth to be legally labelled a "Plymouth Gin"; in fact, in 1933 the distillery won a legal suit against Beefeater producers who had included the word "Plymouth" on one of their labels.

Plymouth's seaside location, along with being the seat of the largest naval base in the country, allowed for its rise as the official gin supplier of the Royal Navy. Indeed, the exclusive contract it enjoyed with the fleet meant that every departing ship received a Plymouth Gin Commissioning Kit: a wooden box containing two bottles of Plymouth Gin and some glasses. Official data confirms that as far back as 1850, the amount of Plymouth Gin supplied was around a thousand barrels a year. The White Star Line also reported its presence on board of their ships in the early 1900s, and one can easily imagine that some bottles of Plymouth Gin sank with White Star's Titanic in 1912. This was a high proof gin, of a higher alcohol content than usual, around 57.15% volume, served in daily doses to the ship's crew. The sailors watched their rations like hawks to be sure they were receiving pure gin and a not watered down ration. They verified its "proof" in a decidedly empirical manner: the gin was poured over gunpowder, and its alcohol content confirmed when a flame was brought close and it ignited. If the powder caught fire, one could be sure of the gin's high proof. The term "Navy Strength" seems to derive from this practice. It remained a popular practice for a long time among the Royal Navy submarine crews: the high alcohol content meant smaller quantities to be stored on board, with greater results. The tradition continued until the advent of nuclear subs, upon which it was strictly prohibited. In time, this type of Plymouth gin disappeared, to be replaced with a 37.5% volume version, considered more competitive on the market. In 1993, however, it saw a rebirth thanks to Desmond Payne, one of

HIGH PROOF AND NAVY STRENGTH

In the USA, 100 degrees proof corresponds to 50 degrees on the Gay-Lussac scale or the Richter alcohol meter, essentially today's volumetric scale whose alcohol density is correctly calibrated at 0.7943. In Anglo-Saxon countries, a different alcohol meter was used until a few years ago, which as a unit of measurement used proof-spirit and was calculated with a Sykes hydrometer, introduced in 1818 and used until 1980. With this scale, 100 proof corresponds to 57.15% in terms of volumetric degrees: in other words, the alcohol content known as "Navy Strength".

the most legendary figures in the world of British gin, at that time the master distiller at Plymouth (today Payne is master distiller at Beefeater, while Sean Harrison has taken over the role at Plymouth).

Even the decision to replace the traditional raw material, corn, with sugarcane molasses did not last long. In 1997, the distillery went back to what is known as the "Old Style", the one-shot distilling method. This technique, different from the two-shot or multi-shot method, consists of placing full alcohol neutral spirits in the still and adding the correct dosage of botanicals, based on the volume capacity of the distiller. After distillation, the alcohol content is lowered with water until the desired strength is obtained. Plymouth Gin has very specific characteristics. For instance, it contains only seven botanicals, which, in decreasing order of importance and weight, are: juniper, coriander seeds, lemon and orange peel, angelica root, orris root, and cardamom seeds. The botanicals are placed in an alembic, in alcohol and water, and left to macerate for the time needed to light the alembic itself. After ninety minutes the boiling point is reached. The entire production cycle lasts just seven hours. The alembic used is considered one of the oldest still working, built by Bennet Sons & Sheras Ltd between 1895 and 1906 (around the same time the Bennet Still was made, of Hendrick's and Tanqueray N°4, and the Beefeater Pot Still), differently shaped from the classic stills, with a much shorter neck and a more bent lye pipe. The distillery houses another two alembics: a Bennet still installed around 1950 and a Carter Head Still built in 1952 and installed around 1960.

Currently the distillery produces some 5,300 liters of gin in every single batch, equivalent to around ten thousand bottles of 41.2% volume gin. Since 1996, bottling has been handled by Hayman's Bottling in Essex. The instantly recognizable bottle features an image of The Mayflower, possibly the most famous symbol of the city of Plymouth but which, for our purposes, is not nearly as famous as its gin.

The classic cocktails: Gimlet, Plymouth Gin Cocktail, Bijou Cocktail, Marguerite Cocktail

GIMLET

TYPE	ALCOHOL CONTENT	TECHNIQUE	GARNISH
SOUR	20.6% abv	Shake & Double Strain	Lemon peel (optional)

RECIPE (8 cl)
4 cl PLYMOUTH GIN
4 cl lime juice cordial
Egg white (optional)

METHOD
This wonderful drink must always be vigorously shaken, yet it can be served without ice in a cold cocktail glass (30 minutes in the fridge should suffice) or in a short tumbler with ice. Fill the shaker almost to the top with ice and stir with a bar spoon until the metal cools. Turn the shaker upside down over a strainer to eliminate any water. Now you are ready to add the ingredients.

First pour in the lime cordial, followed by the gin. Use a jigger for precise doses.

Once all these ingredients have been poured in, add a few drops of egg white (optional). This detail will give the drink an extra soft feeling in the mouth.

Close the shaker firmly and shake in whatever style you choose for at least 10–12 seconds… vigorously! It's crucial that the shaken mixture be "explosive", so that the ingredients blend well, cool properly, and emulsify. Otherwise, you will end up with the opposite effect.

PREPARATION
Home-made lime juice cordial
Squeeze the juice of 3 lemons and 3 limes into a measuring cup, setting the peels aside. Pour the fresh juice into a glass jar.

Sweeten to taste with granulated sugar (usually 1 and ½–2 teaspoons) and dissolve thoroughly in the juice. Add one capful of maraschino (about 2 cl). Remove the pith from the peels and add. Store in the refrigerator overnight.

The next day, taste and adjust for sweetness, adding more sugar as needed. Preparing a cordial in this manner usually requires around 4-5 days.

LONDON DRY GIN

The triumph of botanicals

The definition of London Dry Gin is at once easy to understand and misleading. The term "dry" immediately indicates that this gin is in fact made without sugar, but the reference to London shouldn't be taken to mean a geographical indication. In other words, London Dry Gin can be produced anywhere in the world, even thousands of miles from the chimes of Big Ben. The history of this gin style, likely the most popular among both professional bartenders and enthusiasts today, is relatively recent. Its success stems from the advent of the continuous distillation still in 1830, by an Irishman named Aeneas Coffey, who perfected a method developed by Robert Stein, Scottish, three years earlier. To Coffey's credit—and for which reason stills continue to bear the name "Coffey Stills"—he registered his patent and got the

attention of distillers of his time. One of the first people to note the particular advantages of a continuous still was Alexander Gordon, active in this area going back to 1769, the year he founded his distillery in the London borough of Southwark. His Gordon's Special London Dry Gin was distilled three times, with no added sugar, and aged eighteen months.

Apart from aging, the choice to not add sugar is attributed to the Coffey Still's more effective functioning when compared to pot stills from that era. The spirit obtained from a Coffey still had a higher alcohol concentration and was much purer and cleaner in terms of its aromatics, making the use of sugar almost unnecessary (remember, back then sugar was a means of hiding defects rather than improving a gin's sensory traits). The arrival of Dry Gin was a veritable revolution in the world of distilled spirits, not only because it signaled the start of a style destined for global success: Dry Gin upended the very concept of gin itself, which, when produced without added sugar, allowed the aromatic bouquet of botanicals to rise to the forefront: firstly juniper, followed quickly by all the others. Even Old Tom himself was affected by the introduction of the Coffey Still, abandoning in ever-growing measure its malty notes and evolving into a more herbal and aromatic spirit. From this perspective, Old Tom is considered the link between the Dutch jenever and the modern London Dry, and for a certain period the two co-existed under the labels "sweet" or "dry", with the latter initially meeting the tastes of the upper classes (also on account of the Victorian, more health-focused lifestyle), then branching out to the entire population. Dry Gin's success stemmed not only from its superior elegance and aromatic complexity with respect to Old Tom, but also due to the Coffey Still itself, which provided for large-scale production. This is attested even today by its use in some leading distilleries. Some of the most famous label gins are London Dry Gin, including Beefeater, Tanqueray and Bombay Sapphire. The first opened in 1863, and just thirteen years later included a London Dry in its catalogue (called "James Burrough"). When it finally produced its Beefeater London Dry for the first time, the ensuing success was so great that it became the company's flagship product. Tanqueray, on the other hand, has recently added two other noteworthy products to its classic Tanqueray London Dry Gin, Tanqueray Lovage, characterized by the unusual use of the botanical lovage, and Tanqueray Bloomsbury, taken from a recipe dating to the late 1800s. The same can be said of Bombay Sapphire, whose founder Thomas Dakin is considered one of the leaders of the new generation of distillers. Living during the industrial age allowed Dakin to make use of better technology and the best stills, while also attracting a clientele from among the emerging middle class.

Over time, advanced instruments and deepened knowledge around the art of distillation allowed producers to make

London Dry Gin, by relying on the continuous still. The Coffey Still paved the way for a spirit without defects or bad odors that needed to be hidden, thus one that could be enriched by pleasant notes derived from fruit, spices, roots, and become the sort of "blank page" upon which others could express their own creativity. Provided of course that certain obligatory rules are followed to be able to call the spirit London Dry: no colors may be added, and the botanicals-derived aromas must be natural (so no oils, essences, or dyes) and must be added during the distillation process. And sugar? No more than 0.1 grams per liter. So none, practically.

The only thing left to clarify is the term "London", which, as noted, has nothing to do with geographical location, but recalls instead this style of gin's birthplace. Indeed, several makers benefit from the London Dry label: British gins like Bulldog (which became part of the Campari world in 2014), Caorunn (a gin that maintains its indomitable spirit of Scottish independence by labeling itself a "Scottish Dry Gin"), the English Martin Miller's (which uses water imported from Iceland), as well as the very Italian Ginepraio (a London Dry Gin produced in Tuscany from organic soft wheat grown in the Mugello region). In short, London Dry indicates London only in name, and it has become the gin *par excellence*, the most famous and widespread in the world. Of course, the historical contributions of Dry Gin's "big brothers" remain significant, having "paid for" the greatly improved efficiency in production methods and cleaner flavor. These gins continue to work alongside London Dry even today, rendering the world of distilled spirits even richer and more compelling.

The classic cocktails: Clover Club, Negroni, Pegu Club, Martini Cocktail

Quiet Martini

TYPE	ALCOHOL CONTENT	TECHNIQUE	GARNISH
MARTINI New Era	21.9% abv	Stir&Strain	Pink grapefruit peel and fresh chamomile flowers

RECIPE (9 cl)
4 cl TANQUERAY TEN
1.5 cl Edelflower Elixir
1 cl Timut pepper cordial
2 cl chamomile flower (hard extraction)
A few drops of oleo-saccharum

PREPARATION
Timut pepper cordial
Macerate ten Timuth peppercorns for about 12 hours in 10 cl of sugar syrup (2:1) along with 1 cl of maraschino and 1 cl of vodka/gin of your choice. Taste and filter.

NOTES
The Chilled Reverse Cocktail Martini Glass is a personally "modified" glass that allows for an inverted perspective on the classic Martini glass. The drink is poured into the base rather than the cup vase and is drunk through a side opening.

METHOD
This drink can be made using a mixing glass (often glass but metal is fine) to control dilution owing to temperature maintenance, which for this drink should be very low. Eliminate the excess water. Pour in all the ingredients, being careful to keep the smaller proportioned ingredients for last, as those function as adjusters and as a result need more attention.
Serve in a simple Martini cocktail glass that has been chilled in the freezer or in a Reverse Martini Glass.

MEDITERRANEAN MARTINI

TYPE	ALCOHOL CONTENT	TECHNIQUE	GARNISH
DIRTY MARTINI New Era	36% abv	Stir&Strain	Caper berry rinsed or organic lemon peel

RECIPE (8 cl)
6 cl O'NDINA GIN
0.5 cl Macchia Dry Marino Vermouth
1.5 cl caper kombucha

PREPARATION
Caper kombucha
To begin, make the black tea. While still hot, sweeten the tea with about 100 gr of sugar per liter. Let cool and transfer to a wide-mouthed glass jar. Add some kombucha from another batch or from another mother; the amount should be from 5% to 10%. Next add the SCOBY ("Symbiotic Colony of Bacteria and Yeast"), a gelatinous disk that floats on the surface.

With high summer-like temperatures, fermentation should begin in 10 days. "Cover" the jar with only some gauze, as fermentation is aerobic. Pour 10 cl of this liquid into a small jar and add a teaspoon of salt-packed capers, and let macerate overnight. The following day, filter the liquid, pressing the capers to release their juices.

METHOD
This is a classic Dirty Martini. My advice first and foremost is to store all the ingredients needed for this cocktail in the refrigerator, and better still keep the gin in the freezer. Ideally, chill the cocktail glass in the fridge or freezer, too. Ice the mixing glass well and drain the excess liquid, forcing it a little with the strainer (hard drain). Add the vermouth and gin first, followed by the kombucha. Taste. At this point, you might wish to add or adjust an ingredient to your taste.

SLOE GIN

A gin in name, but not in fact

We'd like to say right off and without mincing words: Sloe Gin is not a type of gin. Rather, it is simply a blackthorn berry (or sloe) liquor made with a gin base. However, for our purposes in this chapter, we will be treating sloe gin in line with common convention. After all, consider that the history of sloe gin begins as a remedy to hide the defects of poor-quality spirits, following the reasoning that if the gin is bad, adding something fruity will surely improve its flavor. At one time there were also Hips or Snag Gin, made with other varieties of blackthorn. The preference for blackthorn, whose precise botanical name is *Prunus Spinosa* (of the Rosaceae family) but is also commonly known to Italians as "strozzapreti" (priest-chokers) or "thorny plum", is likely linked to its availability. At one time, this bush was grown along field borders to indicate property divisions. Moreover, it is a low-maintenance plant, robust against cold and most parasites, adaptable to almost every type of terrain. It's only "defect" is that of being full of thorns, and for this reason gathering sloe berries requires patience... and a pair of good gloves! In recompense, blackthorn is rich in vitamin C, making it a most welcome and beneficial fruit in Northern Europe, a region practically devoid of citrus. Consumption of these berries could take the form of jams or sauces for example, or gin.

The sloe gin tradition was born in England, where the blackthorn bush is plentiful. Defects aside, an alcoholic base with a "fruity" component was particularly pleasing to manual laborers or those who passed several hours outside: alcohol plus sugar meant warmth and energy.

Today several producers are trying their hand at a sloe gin, but not all these produce a classic English variety, choosing Eastern European products instead and in some cases resorting to juices and purées made from the fruit. The alcohol content of sloe gins on the market varies greatly: from 26% vol (Gordon's, Greenall's, Plymouth and Hayman's) to 29% vol (Monkey and Sipsmith), and up to 35% vol (Elephant) and 38% vol (the Italian Rivo Sloe Gin).

Some will bear the word "Damson" on the label. This is a specific variety of blackthorn (*Prunus Insititia*, also called "damascene") originating in Great Britain and, differently from the more commonly used fruits, has a more astringent flavor along with almond notes due to its larger pit.

Sloe gin can be made at home, if you like. To begin, naturally you must first obtain some blackthorn berries, about half a kilo, and possibly of the "frosted" kind (meaning those gathered between October and November). Pit them and place in a bottle. This step is vital, as it is where the hydrogen cyanide comes into play, which is poisonous but also lends a pleasant bitter almond note. Some liquorists hold that, in order to limit the production of cyanohydrin, the fruit should be harvested before the coldest point in the season (hydrogen cyanide increases in cold periods), and the freezing process then concluded in a home freezer, for a much shorter period. This method is both a prevention measure and a means to speed up the harvesting time, which otherwise would be much longer. Once the fruit is placed in the bottle, add around 150 grams of cane sugar, wait a few hours, then fill the bottle with gin. Let rest for at least two months in a cool, well-ventilated place. Then pour out, being sure to filter with a fine mesh strainer. Another way of making sloe gin, one more effective than this cold maceration method (but also more complicated) involves cooking the fruit sous vide for several hours, resulting in a final product with greater acidic notes and bitter finish.

The classic cocktails: Sloe Gin Fizz, Modern Cocktail, The Blackthorn, Millionaire Cocktail I

SLOE_MOTION

TYPE	ALCOHOL CONTENT	TECHNIQUE	GARNISH
COBBLER	13.8% abv	Built	Fresh mint leaves, apple fan and a cherry

RECIPE (20 cl)
3 cl ROBY MARTON GIN
1 cl RIVO SLOE GIN
15 cl apple cider
1 cl Grapefruit & Raspberry shrub
A few drops of Dandelion & Burdock bitter

PREPARATION
Grapefruit & Raspberry shrub
10 cl Grapefruit juice, 5–6 bsp oleo-saccharum, 3 cl raspberry vinegar, 4–5 raspberries. Refrigerate overnight.

METHOD
This drink is made much like a classic Pimm's, directly in a goblet-style glass or a wide water glass. Fill the glass with ice cubes and chill the glass well, being sure to drain off excess water. Carefully pour in all the ingredients except the apple cider, then mix well. The cider is poured in last to preserve some of the carbon dioxide (CO_2). To finish, dust with powdered sugar (optional).

Chapter III

Not by juniper alone...
Botanicals

"The smell of spices evokes places, and I really enjoy the stories told about them"

Moni Ovadia

Perhaps gin's distinguishing characteristic, juniper, also embodies what is most charming and fascinating about this spirit. Yet alongside juniper, an extraordinary array of other botanicals—herbs, fruits, flowers and spices—have lent gin its remarkable distinction over the centuries, engendering not one but many gins, co-existing and simultaneously crafting gin's alchemical beauty. Each and every distiller unfolds this creative potential before us, while alongside them the proficient barman works with a likewise creative mixological spirit, even if preparing a mere "simple" classic like the Gin and Tonic. Depending on the gin, and with a nod to the vital role of the selected tonic, any cocktail glass you hold in your hands will embody these nuanced, diverse emotions. And it is to the hundreds of botanicals, featured in countless recipes, that we owe this wondrous diversity, one that breathes life into every gin and distinguishes one from the next: from the more common botanicals such as cassia, angelica or orris root, to those rarer, more unusual choices like hops, tea, exotic lotus flowers or Buddha's Hand. Very often these botanicals tell the stories of their place of origin, expressing in the gin itself, enchantingly, the nature and qualities of their respective lands.

We therefore felt it important to explore the subject of botanicals with the following dedicated chapter. While recounting all of them would have been nearly impossible, here we are pleased to present a worthy selection of over forty plants, flowers, seeds and roots. Together they form a realistic picture of what truly lies behind the word "gin", a word we are all familiar with but perhaps have known, until now, only superficially.

Ambrosia Wine Cobbler

TYPE	ALCOHOL CONTENT	TECHNIQUE	GARNISH
COBBLER Sparkling	12.8% abv	Built and Twist'n Sparkle	Mint leaves, citrus fruit slices, red berries and powdered sugar

RECIPE (20 cl)
10 cl oxidized ambrosia wine
3 cl PANAREA GIN
Thinned with Scortese Ginger Beer

METHOD
Pour the oxidized wine and the gin into a Twist'n Sparkle carbonator. It's vital that all the ingredients are very cold. Close firmly and carbonate. Let rest for a few minutes in the refrigerator to allow the gas to combine well with the liquid. Meanwhile, fill a goblet or nice wine glass with crushed ice and pour in the wine that's been fortified with gin. Finish with the ginger beer and garnish to taste.

PREPARATION
Ambrosia wine
In an uncovered jar, macerate some aromatic white wine (similar to Prosecco, IM6013, Riesling, etc) with 2 cls of honey mix, rosemary, calamint, thyme, lemon and orange peel, pepper and star anise. Refrigerate for a few days.

SAGE
Salvia Officinalis

Distribution: Central Europe and Mediterranean maquis.

Propagation: cutting.

Parts used: leaves.

Harvest: leaves, all year round.

Proprieties: used for a range of female reproductive ailments, such as premenstrual syndrome and menopausal side effects; helps with menstrual flow in cases of amenorrhea, and, given its antispasmodic properties, is used for intestinal tract diseases as a smooth muscle relaxant. Useful in the treatment of intestinal irritability, digestive tract spasms, and menstrual pain. Moreover, its attributes include anti-inflammatory and diuretic properties, as well as being antiseptic and soothing.

Curiosity: the word "sage" ("salvia" in Italian) derives from the Latin *salus*, meaning "health", or from *salvus*, "safe" or "protected". Sage contains a complex ketone called thujone, which in high doses can be toxic.

Notes: herbaceous and fresh.

LAVENDER
Lavandula Angustifolia Miller

Distribution: Mediterranean region.

Propagation: pollination.

Parts used: flowers.

Harvest: starting with two-year-old plants; ideal harvest time is when the flowers just begin to bloom, between July and August.

Proprieties: lavender has been known for various properties since ancient times, including antiemetic, antiseptic, analgesic, anti-bacterial, vasodilative, antineuralgic, and as a treatment for muscle pain. It is also considered a mild sedative.

Curiosity: the word derives from the Latin gerund of the verb "lavare" (lavandum = that which must be washed), an allusion to this plant's frequent use in antiquity, the medieval period especially, as a means to wash the body.

Notes: lavender is a widely used plant that bestows notable freshness and soft floral notes. It helps to extend the retro-nasal component and bind different spices together.

ROSEMARY
Rosmarinus Officinalis L.

Distribution: Mediterranean region.

Propagation: cutting.

Parts used: sprigs.

Harvest: all year.

Proprieties: antiseptic, excellent for treating excessive triglycerides in the blood, for liver and gallbladder problems, dyspepsia and intestinal bloating. It is an effective remedy against phlegm and gastrointestinal problems.

Curiosity: its name, *ros marinus*, means "sea dew".

ALEXANDRIAN SENNA OR CASSIA
Cassia Angustifolia

Distribution: this plant grows in warm, dry climates, and originated in North Africa and the Middle East.

Parts used: bark.

Harvest: from March to September.

Proprieties: cassia leaves produce a laxative effect, given that they stimulate contraction of the intestinal walls, re-balancing the two types of colon contraction.

Notes: very similar to cinnamon, with a high oil content that lends sweet and warm notes.

LONGAN OR DRAGON'S EYE
Dimocarpus Longan

Distribution: tropical belt of Southeast Asia.

Propagation: seeds, cuttings.

Parts used: fruit.

Proprieties: contains high levels of vitamin C, potassium, phosphorus and copper. Antioxidant, anti-aging, analgesic and anti-inflammatory.

Notes: lends sweet fruity notes and softness on the palate.

FENNEL
Foenicularum Vulgare

Distribution: Mediterranean maquis.

Propagation: seeds.

Parts used: flowers, seeds, umbels.

Harvest: spring/summer.

Proprieties: stimulates appetite, improves digestion and helps dyspepsia, aids hiccoughs and halitosis, and also has diuretic properties.

Curiosity: in Greek, the word marathon is the name of the Hellenic location where the Athenians defeated the Persians in 490 B.C., and it literally means "field of fennel".

In medieval magic, fennel was used to protect against deception and witchcraft.

Notes: lends fresh notes, and expands flavor perception within the botanical range in which it is used.

GERANIUM
Geraniaceae or Cranesbillis

Distribution: originally from Southern Africa, now cultivated in much of Europe.

Propagation: seeds.

Parts used: flowers and leaves.

Harvest: July/August.

Proprieties: used in aromatherapy for its fundamental role in balancing the nervous system. Antidepressant, anti-inflammatory, calming, astringent and antiseptic. Geranium stimulates the lymphatic system and is an effective tonic for the kidneys and liver.

Curiosity: from the Greek *géranos*, meaning "crane", for the shape of its fruit. With some 160 species in this genus, plus many hybrids, it is difficult to make any universal statements about geranium. It was introduced to Italy by a Venetian.

Notes: geranium pelargonium, which is used in Geranium Gin, contains essential oils such as geraniol (present also in mango, apple, raspberry, melon, and carrot), linalool (iris, lily, lavender, cardamom, cinnamon, and cassia), citronellol (lemon, orange, mandarin, grapefruit, and pomegranate), borneol (coriander and liquorice), rose oxide (rose, ginger, elderberry, and eucalyptus) and menthone (mint).

BUDDHA'S HAND
Citrus Medica Sarcodactylis

Distribution: Eastern and Southern Asia.

Parts used: fruit.

Harvest: the main flowering is in spring, but it flowers well throughout the year, and as a result the plant often bears ripe and unripe fruits (large and small) and flowers at the same time.

Proprieties: the fiber-rich peel is beneficial to intestinal function. Moreover, it contains flavonoids, powerful antioxidants for the entire body, and is considered diet-friendly, low in calories.

Curiosity: the fruit segments divide early in the development to form several individual fruits, and these segmented outgrowths account for the fruit's non-spherical, hand-like shape, leading to a common perception of this fruit in The East as "the hand of Buddha".

In Japan, this citron is considered a good luck fruit and is usually gifted at New Year. It is also used to perfume linens and clothes when placed in drawers and wardrobes.

Notes: this particular variety of citron is the result of a genetic deformity. Buddha's Hand is actually an ancient mutation of citron, one of the oldest types of citrus, native to the foothill valleys of the Himalayas in India, on the border with Bhutan.

GRAINS OF PARADISE
Aframomum Melegueta

Distribution: native to West Africa's coastal countries: Ghana, Liberia, Ivory Coast, Togo, and Nigeria.

Parts used: seeds.

Harvest: all year.

Proprieties: numerous clinical studies are currently underway to evaluate this plant's possible anti-inflammatory and metabolic activator pharmacological properties.

Curiosity: this herbaceous, perennial tropical plant is part of the ginger family.

Notes: in Congo, the seeds are used in ritual magic, as bracelets containing seven (or multiples of seven) grains.

LIQUORICE
Glycyrrhiza Glabra

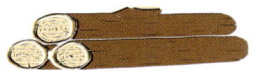

Distribution: native to Southwestern Asia and Mediterranean regions.

Propagation: principally by division of the adult plants.

Parts used: rhizomes, roots, stolons.

Harvest: perennial herbaceous plant.

Proprieties: the main active agent in liquorice is glycyrrhizin, which is anti-inflammatory and anti-viral. Current research is focusing on these beneficial properties and how to identify their uses from a healing perspective: in the treatment of ulcers, chronic liver diseases, and the prevention of serious auto-immune diseases. It is an excellent emollient, and fluidifying in the treatment of bronchitis, colds, flus, coughing fits, and problems with the mouth and oral cavity. It is also effective as a renal and intestinal antispasmodic and is excellent in the treatment of gastric and duodenal ulcers. It contains laxative and appetite-suppressant properties, helping reduce appetite and increasing a feeling of fullness. Moreover, it functions as a mineralocorticoid in the treatment of Addison's disease. Used externally, it has a soothing, balancing effect.

Curiosity: the roots are treated in chemical laboratories, primarily on an industrial scale, to produce a paste. This is then made into "black liquorice" and sold in the form of sticks, candies and drops.

Notes: thirst-quenching, refreshing, flavoring, aromatizing and sweetening. The glycyrrhizin found in the roots is in fact sweeter than sugar and can be given to diabetics.

ITALIAN HONEYSUCKLE OR WOODBINE
Leonicera Caprifolium L.

Distribution: Europe.

Propagation: climbing plant.

Parts used: flowers.

Harvest: May/July.

Proprieties: it is believed to aid those people who regret missed chances, who suffer from a sense of time passing too quickly, who hang on to old loves, are held fast by memories, or reject new experiences.

Curiosity: the Latin name of this plant, *caprifolium*, likely indicates the "passion" that goats have for its leaves.

Notes: lends interesting warm floral notes and olfactory softness.

CLOVE
Syzygium Aromaticum

Distribution: warm regions of Asia and America.

Propagation: evergreen tree.

Parts used: flower buds.

Harvest: before the flower closes.

Proprieties: energizing and antiseptic, antineuralgic, antispasmodic, clove is good for the stomach and the intestine, and is very useful in the treatment of colds.

Curiosity: in ancient China, around 200 A.D., whoever spent time near the emperor was obliged to chew cloves, to perfume and purify their breath.

In the past, cloves have been greatly used to treat nausea and vomiting, along with coughs, infertility, and as an analgesic for tooth aches.

Notes: long-lasting warm notes and a delicate piquancy.

OLIVE
Olea Europea

Distribution: Mediterranean basin.

Propagation: cuttings.

Parts used: leaves and fruits.

Harvest: from November to February.

Proprieties: peripheral vasodilator, diuretic, very effective in the treatment of hypertension. Excellent also for the treatment of colic (biliary and renal) and stones (bile duct and kidney). Olive oil has noted benefits for burns.

Curiosity: according to legend, Athena and Poseidon competed for the right to name the capital of Attica. A contest was proposed to resolve the matter: which of the two deities could give the city the most useful item would win the right to baptize the city. Poseidon gave birth to a beautiful steed, while Athena brought a tree brimming with olives to life... and the rest, as they say, is history—till this day, winners of competitions are still crowned with olive wreaths. Also worth noting: athletes would spread olive oil on their bodies before competitions or battle.

Notes: the Bible tells us that the dove Noah released after the great flood returned to the ark carrying an olive branch in its beak, a sign of heavenly blessings and peace.

ROSE HIPS
Rosa Canina L.

Distribution: throughout Europe.

Propagation: cuttings.

Parts used: leaves, flowers, and fruits.

Harvest: summer/autumn.

Proprieties: the leaves are used to treat diarrhea and kidney and bladder stones. The petals are refreshing, laxative and effective for sore throats.

Curiosity: the name of this plant derives from what ancient peoples held to be its most important function—as an excellent remedy for dog rabies—though subsequently it became known as a treatment for all diseases. It was for this reason Swedish naturalist Linnaeus, father of modern botany and zoology, named this rose "canine" in the 1700s. Apparently, too, the crown of thorns placed on Jesus Christ was made with rose hip branches.

The fruit is rich in vitamin C (100 gr of rose hips contain the same amount as 1 kg of oranges). In fact, during the Second World War, the English gave their children rose hips in place of citrus.

Notes: its scent has revitalizing properties in cases of psycho-physical stress. It noticeably opens up the floral olfactory spectrum and lends light astringent notes to flavor.

THYME
Thymus Mongolicus – Thymus Vulgaris L.

Distribution: from sea level to 1,500 meters altitude.

Propagation: dividing the plant head.

Parts used: entire plant.

Harvest: March/April.

Proprieties: antiseptic and disinfectant. Specifically indicated for upper-respiratory problems, coughs, pertussis (whooping cough), and acute and chronic bronchitis.

Curiosity: its name derives from the Greek verb *Thyo*, meaning "to make sacrifices". The ancients burned it during their ritual offerings to their gods, on account of its distinctive, penetrating scent.

Notes: given its antiseptic properties, thyme has been widely used in the preserving of meat, as well as food preparation. It was used in decoctions and infusions, as well as creams and balms. In the last century, French chemist Lallemand managed to extract its essential oil, which he named "thymol", subsequently using it as an antibiotic.

CHAMOMILE
Chamaemelum Nobile

Distribution: Europe.

Propagation: pollination, seeding.

Parts used: flowers.

Harvest: July/September.

Proprieties: sedative, antispasmodic, anti-inflammatory, antiseptic and antimicrobial. Frequently used to treat headaches, insomnia, anxiety, toothaches, and muscular and menstrual pain. It relieves sunburn, dermatitis, and red or dry skin.

Notes: lends wonderful aromatic depth and a touch of softness when used in alcohol.

STAR ANISE
Illicium Verum

Distribution: East Asia.

Propagation: seed.

Parts used: the fruit (inaccurately called "seeds") are gathered when still green, then left to dry in the sun, turning a reddish-brown color.

Harvest: tropical evergreen tree.

Proprieties: star anise is used as an antispasmodic, stimulant and stomachic. Its main properties include digestive, carminative, anti-diarrhea, and galactagogue. It's also used as a stimulant of exocrine glands.

Curiosity: in 1558 the English explorer Sir Thomas Cavendish brought star anise fruits to Europe that he collected during his travels.

Thus, the first place this fruit landed on the Old Continent was in London.

Not long after, it made its way into the hands of the famous and esteemed court pharmacist Hugo Morgan, and then to Jacob Garot, from whom Clusius was then able to obtain a sample, which he described in highly analytical detail in 1601.

Notes: the term "Illicium" derives from the Latin *illicio*, meaning "to attract", while in Persian, star anise is called "bādiyān", from which we get the French name "badiane".

The edible part is made of anethole, a scented essential oil found in the seeds.

GINGER
Zingiber Officinale Roscoe

Distribution: native to India and Malaysia, cultivated in tropical areas.

Propagation: underground rhizomes.

Parts used: rhizomes.

Proprieties: useful in treating intestinal gas, anorexia and fever, and indicated for gastric atony as it stimulates enzymatic secretion and balances bacterial flora.

Curiosity: at the end of banquets, a kind of ginger bread would be served to guests to aid digestion. Over time, this well-known "gingerbread" grew to become a kind of dessert and later, in the Anglo-Saxon tradition, the bread was made in the shape of a person: hence, the birth of the gingerbread man cookie.

Notes: ginger is believed to ward off evil spirits. In distillates, ginger lends a long finish, and highly interesting spicy and peppery notes.

CARDAMOM
Elettaria Cardamomum

Distribution: native to the rain forests of Kerala (in the southern Indian subcontinent) and Sri Lanka, today cardamom is also successfully cultivated in Guatemala in Central America.

Propagation: division of the rhizomes.

Parts used: seeds.

Proprieties: cholagogue and choleretic, cardamom aids the digestive process, has antibacterial and antifungal properties, and counteracts dyspeptic problems. Widely used in Southeast Asian countries to treat halitosis, tooth infections and gingivitis, and in the prevention of throat diseases, pulmonary congestion and stones.

Curiosity: the term "Elettaria" derives from the Tamil language, meaning "leaf granules". For ancient Romans, cardamom was among the most expensive spices in the empire, together with vanilla and saffron. In the 1800s it was considered an effective sexual stimulant, thus gaining fame as a powerful aphrodisiac.

Notes: cardamom is one of the most popular botanicals used in the world of gin, for both its highly delicate aromas in the post-distillation phase as well as its prized invigorating notes.

RASPBERRY
Rubus Idaeus L.

Distribution: Central and Northern Europe; widespread in the Alps and Apennines.

Propagation: division of the basal shoots (the new shoots that grow from the base of the plant).

Parts used: leaves and fruits.

Harvest: summer.

Proprieties: used for rheumatism, gout and arthritis, raspberry is also very effective in treating oral and throat inflammation.

Curiosity: according to the original myth of Mount Ida, the Great Goddess "Rhea" (or "Cybele"), mother of all the Olympians, gave birth to Zeus and hid him from the baleful glances of his father, Cronos. On this same mountain, Venus seduced Anchises and Mercury, to produce, respectively, Aeneas and Eros.

Notes: highly acidic and fresh, it stimulates the flow of saliva.

ALMOND
Amygdalus Amara

Distribution: native to China and now cultivated throughout the Mediterranean belt.

Propagation: seeding and subsequent grafting.

Parts used: fruit.

Harvest: August/September.

Proprieties: bitter almonds are sedative and antispasmodic, while the sweet ones are purgative and emollient.

Curiosity: Prunus dulcis is a fruit tree belonging to the Rosaceae family and the genus prunus. Almonds are the seeds of this tree.

In ancient times, almond trees were considered a remedy for drunkenness. Plutarch tells the story of a certain doctor who would challenge anyone to a wine-drinking contest, and who always came out the winner. One day, however, he was discovered eating bitter almonds before the contest and was obliged to confess that without the benefit of eating almonds, even a small amount of wine would have gone straight to his head.

Notes: most almonds are sweet, but some bitter ones are cultivated. These seeds are considered highly toxic because they contain amygdalin, which can cause cyanide poisoning. If digested in large quantities, they can bring on headaches, vomiting and, in extreme cases, death, especially in children.

ELDERBERRY
Sambucus Niger

Distribution: up to 1,200 meters in forests and on river banks.

Propagation: cuttings.

Parts used: leaves and flowers.

Harvest: August/September.

Proprieties: galactagogue, antirheumatic, antispasmodic, emollient and sudoriferous. Effective in treating coughs, colds, bronchitis, flus, fevers, and infectious diseases involving skin rashes. Antiarthritic, antirheumatic, anti-gout, purgative.

Curiosity: the origins of this plant's name derive from how its branches were once used in ancient times: to make a triangular-shaped string instrument similar to a small harp that the Greeks called "sambice" and the Romans called "sambuca".

Notes: the flowers are known for their ability to open the aromatic spectrum and to act as a binding agent between other used spices.

COMMON HOPS
Humulus Lupulus L.

Propagation: pollen division.

Parts used: the female inflorescences and the glands that cover the bracts.

Harvest: August/September.

Proprieties: known for its depurative and laxative properties.

Curiosity: hops is widely used for its organoleptic qualities, but mostly for its remarkable preservative properties. It was also held to be an effective sleep aid and thus was used to stuff pillows.

Notes: it is difficult to describe in a few words what the "simple" hops can bring out in a gin. Each of its dozens and dozens of different qualities lends a different, highly interesting quality to the distillate. Here we will attempt to summarize them succinctly while not detracting from the plant's significance in the production of gin. Its decidedly bitter note, contrary to what one might think, lends length to the finish and freshness when well-handled in the extraction phase, as well as, sometimes, tropical, citrus and floral notes.

PEPPER
Piper Nigrum

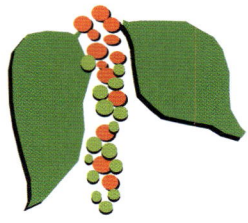

Distribution: Indonesian archipelago, Asia and Equatorial America.

Propagation: cuttings; the plants flower after 2–3 years and reach their maximum productivity after 3–4 years.

Parts used: fruit.

Harvest: when ripe.

Proprieties: antiseptic.

Curiosity: a well known stimulant and aphrodisiac.

The word "pepper" derives from the Latin *piper*.

Notes:

Black pepper – is produced from the unripe fruit of the pepper plant. The fruits are briefly blanched in hot water to both wash them and prepare them for drying, when the breaking of the pulp speeds up the blackening process. The grains are dried in the sun or by using dedicated dryers for several days, during which the fruits dehydrate and blacken.

White pepper – is made by soaking the pepper berries in a bath for about a week. The pulp decomposes and can be easily removed, then the mesocarp is removed, and the seed is dried.

Green pepper – as with black pepper, is produced from the unripe fruit. It is treated with sulfur dioxide during the drying process to preserve the green color.

Cubeb pepper – these are berries of the Piper cubeb plant, also known as "Java Pepper". It comes from Indonesia and is grown on the islands of Java and Sumatra, in the Malay Archipelago, in the Antilles and some parts of Africa (such as Sierra Leone and Congo).

Long pepper – Piper longum, pimento, or "Bengali pepper".

Allspice – a spice erroneously linked to pepper due to another of its names, pimento (which means "pepper" in Spanish). The name "pimento" derives from the French piment, from the Latin *pigmentum*, meaning "pigment" or "color". In English it's known as "Allspice", on account of its seeming blend of cinnamon, nutmeg, and clove.

Sichuan pepper – the berries of an Asian plant of the genus zanthoxylum, native to its same-named province in China.

Timut pepper – native to Nepal, and belonging to the same family as Sichuan pepper, timut is not actually a pepper but rather the skin of a berry.

CINNAMON
Cinnamomum Zeylanicum Nees

Distribution: native to Ceylon and Malaysia.

Propagation: seeds and cuttings.

Parts used: inner bark.

Proprieties: toning and stimulating, cinnamon is also an intestinal and pulmonary disinfectant.

Curiosity: along with pepper, cinnamon was one of the most prized and sought-after spices during the Roman Empire and at the time of Marco Polo's voyages to China.

The name comes from "cinnamomo", which in its day came from the Arab Kinnamon: literally, "painful". The Chinese considered cinnamon one of the oldest spices in the world, having used it in their country some 2,700 years before Christ.

Notes: a very warm and enveloping spice that lends sweetness and softness to gin.

LEMON BALM
Melissa Officinalis L.

Distribution: widespread.

Propagation: seeds or by dividing the plant head.

Parts used: leaves and flowering tips.

Harvest: spring/summer.

Proprieties: used to treat nervous disorders, depression, melancholy and hypochondria, as well as stomach and intestinal disorders.

Curiosity: the word Melissa derives from the Indo-European root mel, from which we have terms such as "miele" (honey in Italian) and "idromele" (mead in Italian): in short, sweetness. Honey has long been symbolically linked to wisdom and immortality. In Greek, melissa means bee, an insect highly regarded for its presumed connection to the sun. Moreover, various miraculous properties are attributed to it, such as lifting the spirit, countering depression and nervousness, and easing melancholy.

Notes: lemon balm's scent is highly calming and is tonic for stress. In distillates, it lends softness (what some also call "sweetness") along with marvelous floral notes with hints of citrus.

CORIANDER
Coriandrum Sativum

Distribution: throughout Southern Europe.

Propagation: seeding.

Parts used: fruit.

Harvest: July.

Proprieties: stimulates gastric functions and aids rheumatic pain.

Curiosity: the name derives from the Greek Koris, meaning "stink bug", thought to be attributable to the plant's smell when in a vegetative state, similar to that of the insect. It is also called "Chinese parsley".

Notes: in Europe, starting in the 16th century, sugar-covered coriander seeds became a much sought-after candy, the type Italians call "confetti", and subsequently the name of this plant came to indicate the decorative paper pieces associated with Carnival, called "coriandoli" in Italian.

WINTER SAVORY
Satureja Montana L.

Distribution: Asia and Europe.

Propagation: division of the plant heads and cuttings.

Parts used: leaves and flowering sprigs.

Harvest: spring/summer.

Proprieties: used to treat dyspepsia, digestion problems, vomiting, diarrhea, intestinal gas and stomach ache.

Curiosity: since ancient times, it has been considered a highly potent aphrodisiac, so much so that it was banned in monasteries. The Italian name for this plant, "santoreggia", derives from "Satiro", the mythical woodland creature with the half-man, half-goat appearance whose most noteworthy trait was that of insatiable lust. According to others, however, the word derives from the Latin *saturejum*, at that time related to the word "satura", meaning "sauce" or "mixture", since this herb was mixed into everything: a fact reflected in its English name as well, "savory".

Notes: an aromatic herb frequently used in cooking, in particular to flavor meat dishes or salads. This strongly Mediterranean characteristic, the marvelous balsamic and herbaceous notes present, render savory a fundamental ingredient for those seeking a "Provencal-like" note in the spice compounds.

CUCUMBER
Cucumis Sativus L.

Distribution: native to India.

Propagation: seeding.

Parts used: fruit.

Harvest: during summer.

Proprieties: excellent for hydration and skin care, but not easy to digest, cucumber fruit is prized for its anti-constipation effects, while the juice is purifying and refreshing. An effective aid against urinary tract inflammation and against bronchial catarrh.

Curiosity: it is mentioned in the Bible as a delicacy and for its fundamental role in the diet.

Notes: the undisputed symbol of summer cooking, this vegetable possesses a surprising range of aromatic possibilities. It greatly increases fresh notes, especially on the olfactory level, but also, on the gustative level, lends elegant, thirst-quenching vegetable notes.

PEPPERMINT
Mentha Piperita L.

Distribution: widespread.

Propagation: dividing the plant head, seeding.

Parts used: leaves and flowering tips.

Harvest: summer.

Proprieties: it has applications for intestinal bloating, spasms, gastritis, diarrhea, and against nervousness, insomnia, hysteria and menstrual pain.

Curiosity: peppermint was cataloged for the first time in 1696 by English botanist John Ray who, during the course of his work classifying the various types of mint, discovered this exemplar's more intense smell, distinct from other mints. He called it "peppermint".

Notes: the name "mint" is associated with the nymph Minthe, who in Greek mythology had the misfortune of being so beautiful as to make Hades, god of the underworld, fall in love with her. For her sake, Hades began to neglect his wife, Persephone, who in turn took revenge on the nymph by transforming her into a plant. Hades could not bring Minthe back to life; as a final gesture of love, however, he bestowed upon her the characteristically fresh scent we all know.

SAFFRON
Crocus Sativus L.

Distribution: native to Asia Minor, it also grows in the Mediterranean basin.

Propagation: bulbs.

Parts used: stamens.

Harvest: October/November.

Proprieties: useful for treating dyspepsia, insomnia and hysteria.

Curiosity: it takes some 200,000 flowers and 500 hours of labor to obtain 1 kilogram of saffron, facts that explain its high price.

A substitute plant called safflower, also known as "false saffron", is often mixed with it to bring costs down.

Notes: commonly referred to as "red gold" on account of its weight-to-price ratio, saffron has a distinctive scent, with slightly bitter notes and earthy/warm hints, accompanied by notes of honey and rose that awaken a floral sweetness. Those who use it usually work it at the end of the distillation process, to take advantage of the delicate color notes as well.

BASIL
Ocimum Basilicum L.

Distribution: Mediterranean maquis and all temperate zones.

Propagation: seeding.

Parts used: leaves and flowering tips.

Harvest: spring/summer.

Proprieties: stimulating and beneficial for the nervous, circulatory and muscular systems. In addition to digestive and anti-fermenting properties, basil is also antispasmodic and is an effective expectorant.

Curiosity: basil is native to India.

Etymology: the name derives from "Basilisk" (a serpent who could kill with a glance, and against which the plant was believed to be an antidote) or, as others hold, from the Greek basilikòs, meaning "regal".

Notes: according to legend, the Empress Helen, mother of Constantine, found basil on the site of Christ's crucifixion.

TEA
Camellia Sinensis

Distribution: native to China and Japan, it is produced throughout Southeast Asia.

Parts used: leaves.

Proprieties: antiradical and antioxidant. A cup of green tea provides the active equivalent of five portions of raw fruit or vegetables, or 400 mg of vitamin C. It has antioxidant effects on LDL ("bad") cholesterol and is useful in the prevention of arteriosclerosis. Positive effects on vascular function have also been noted.

Curiosity: China produces thousands of varieties of tea, of which about 50% is green tea. The remaining 50% is instead comprised of red tea, also commonly known as "black" tea.

Notes: during production, tea is a delicate element to work with, decidedly difficult during extraction, given that some of its aromas present lastly, as tail notes. Unlike many other botanicals, however, tea works very well throughout the various phases, in terms of its palatal taste, its strengthening of vegetable notes in the cleaning phase, and its degreasing function in the oral cavity. It is particularly noticeable, too, in the retro-olfactory phase, extending the olfactory length of the aromatic spectrum of the distillate in question (albeit in a delicate way).

NUTMEG
Myristica Fragrans

Distribution: a tree from the island of Banda in the Maluku islands (Indonesia), nutmeg is today found in tropical regions.

Propagation: pollination by small beetles called Formicomus braminus.

Parts used: seeds and exocarp.

Harvest: production begins after 7–8 years and reaches full production after 20. This plant can be propagated by seed.

Proprieties: when dissolved in water and ingested in high doses (10–20 gr), nutmeg causes a slight alteration of consciousness and likely visual hallucinations, owing to the presence of myristic acid. This has earned it the nickname "the poorman's narcotic". Once considered a cure for more than one hundred diseases, nutmeg today is known to alleviate acute pain, such as rheumatic and muscular.

It contributes to relieving the symptoms of the common cold, and is anti-inflammatory, antibiotic and expectorant. Though useful in treating nausea, dizziness, and dysentery, it must be taken with extreme caution, because an overdose can be deadly.

Curiosity: nutmeg is the seed, while the external part that covers the seed is the macis.

During the Gold Rush, nutmeg was so valuable that its weight was equal in value to gold. An anecdote from the time speaks to the spice's incredibly high value: apparently a sailor was able to set himself up for life, after having taken a trip to countries where nutmeg was produced and managing to steal a few that he then sold on the black market. Apparently several such expeditions have been made, with the sole purpose of finding routes to the precious nutmeg and bringing it back to market.

Notes: nutmeg's scent is almost unmistakable. It lends a warm flavor to gin, at once marvelously earthy and slightly spicy, along with a notable aromatic complexity, elegance and sophistication.

CITRUS

LEMON
Citrus Medica L.

Native to Asia, lemons were introduced to the Mediterranean basin around 1200 by the Arabs. Italy is one of its first large-scale producers, in Sicily, Calabria and Campania. It is propagated by cuttings or seeds, and both the flowers and fruits are used. In the Arab world, lemons were widely used to treat poisoning and bites from poisonous animals, and for centuries has been considered a cure-all for contagious diseases: for example, a lemon pierced with cloves in the shape of a cross would be placed in the hand of those who'd died from the plague. Lemons have effective thirst-quenching, astringent and slimming qualities, as well as exceptional anti-scurvy and antibacterial properties. Excellent for treating nausea, stomach ache and feverish illnesses.

ORANGE/BITTER ORANGE
Citrus Sinensis/Citrus Bigaradia Loisel

Native to India, today oranges are also grown in Mediterranean regions. They are propagated by seed. The parts used are the leaves, flowers and peel. Oranges are harvested in winter (the flowers, however, in April/May). The word "orange" derives from the Arabic *narang*, which in turn comes from the Sanskrit *naranjia*, meaning, literally, "the elephant's favorite fruit". It has tonic, eupeptic and febrifugal properties.

BERGAMOT ORANGE
Citrus Bergamia

The name comes from the Turkish *Beg Armudi*, "the lord's pear" (its shape is very similar to that of the bergamot pear). The first intensive orchard producing the bergamot orange was established by Nicola Parisi along the coast of Reggio Calabria in 1750. The flowers bloom in spring, while the harvest takes place in June/July. To obtain 1 kilogram of essence 200 kilograms of fruit are required. Some 80% of global production of bergamot comes from Calabria. It is a fundamental ingredient in Eau de Cologne.

YUZU
Citrus Junos

Found in East Asia and Tibet, this is a hybrid of mandarin orange and the lemon-like Ichang papeda. One of the areas best suited to its cultivation is Shikoku, located on the inland sea of Seta, an island of the Japanese archipelago considered the nation's most unspoiled natural area. Japanese warriors have used yuzu since ancient times to restore their energy and strengthen the immune system, thereby avoiding infections.

POMELO
Citrus Maxima

Also commonly called "pummelo" or "pampaleone", pomelo is considered by some the oldest citrus cultivated by humans. It is one of three species from which all today's citrus fruits derive. Native to Southeast Asia and Malaysia, pomelo is the largest of all citrus, with fruits weighing up to nearly 2 kilograms. Along with citron, mandarin, and papeda, it is one of the four non-hybridized species.

LIME
Citrus Aurantifolia

Also known as "lima" or "limetta", it is widespread in Malaysia and India, but today is cultivated also in Mexico, Latin America, Southeast Asia, the Caribbean and California, from where it is exported throughout the world for its antiscorbutic qualities. Even today one of London's docks is still called "Limehouse", recalling the warehouses that stored this fruit when it was taken off arriving ships. More than 160 different types are found around the world, but the two most common varieties are the Mexican lime and the Persian lime.

MAKRUT LIME
Citrus Hystrix

This is thought to be a hybrid of citron and citrus limetta. Native to Southeast Asia, it is grown in Thailand, Vietnam, Laos and Cambodia, yet is also found in Indonesia and Madagascar. The leaves can grow very long (up to 12 cm) and are extremely aromatic. The fruit, rich in vitamin C and antioxidants, boasts anti-inflammatory properties and is suitable for those suffering from sore throat and asthma. It is also effective against cancer, thanks to the presence of limonin.

JUNIPER
Juniperus Communis

Distribution: throughout the Mediterranean and mountain belts.

Propagation: seeds and cuttings.

Parts used: fruits and leaves.

Harvest: the fruits in autumn.

Proprieties: juniper has been officially approved only for the treatment of dyspepsia, given the carminative and stomachic effects carried out by its essential oil and also, most likely, on account of the resinous substances found in the plant's fruit. Yet many other properties are also attributed to juniper: diuretic, anti-inflammatory, hypoglycemic, hypotensive, antiseptic, and antiviral against the Herpes simplex virus. It is also used to treat asthma and bronchial catarrh, and to relieve stomach acid.

Curiosity: juniper was prized in ancient times for its presumed magical properties. Greeks and Latins would burn it during ritual ceremonies as a symbol of health and fertility. During the Middle Ages, juniper bunches were sewn into clothes as talismans. Other customs included hanging it in stalls to protect animals and on house doors to ward off witches. According to popular belief, burning juniper would aid against infectious diseases as it helped to purify the air, a practice that was still in use in hospitals during the Second World War.

Notes: the genus juniperus includes more than fifty different species. Its distribution is virtually world-wide, with juniper present on all five continents.

The berry, actually a cone, is similar to a pine cone. Its woody scales form as the cone matures, taking on its typical compact appearance. Ripening varies from 18 to 36 months. Almost all the existing varieties are ornamental, being slightly toxic to humans. Among the "edible" species, however, in addition to the Communis we have the Occidentalis, used in the Americas.

The same plant, grown in different parts of the world, brings paradoxically opposing results. According to research carried out just a few years ago, the Greek Juniperus Communis contains much less alpha-pinene (the typical molecule of pine notes) than other European types. Those grown in Montenegro and Iran, on the other hand, contain almost twice the amount. Another precious molecule contained in juniper is limonene, responsible for its soury-orange scent. In short, depending on the terroir, juniper can lend differing notes to gin. In his beautiful 2012 work *Gin: The Art and Craft of the Artisan Revival in 300 Distillations*, Aaron Knoll discusses the results of a study on mono-origins, creating a series of gins called "Origin", each with a different juniper as the sole ingredient.

Here we will describe some of these "territorial" varieties:

Skopje, Macedonia:

An unripe juniper, slightly resinous on the nose. Black peppercorn notes and lively spicy hints. Spiced and warm on the palate, with an initial unripe note. Mid-palate, a resinous aroma. With the finish, a hint of sea salt and almond.

Meppel, The Netherlands:

Earthy, with a typical woody retro-olfactory note. On the palate, light mint and herb notes with hints of rosemary. Herbaceous finish with hints of almond.

Arezzo, Italy:

Resinous and woody, the juniper is very concentrated in the top notes, which immediately fade and give way to woody hints and an intense orange peel aroma. Very classic on the palate: it opens with a touch of citrus, vanilla and lively unripe juniper. The finish has notes of vanillin and orange peel, with an explosion of warmth.

Istog, Kosovo:

Unripe juniper tending to pine, with violets and coriander on the nose. On the palate, it is surprisingly sweet and floral. Cherries, then a slower juniper tending to pine, invigorating and sharp. The finish has hints of almond and a lasting intense juniper tartness.

Veliki Preslav, Bulgaria:

On the nose, notes immediately recall coriander, a slightly more subtle juniper, and floral notes as well. Opens fruity on the palate but with an intense herbaceous juniper at mid palate. A complex floral edge, with hints of coriander and hibiscus. Long and lively finish, dominated by juniper.

Klanac, Croatia:

On the nose, a slightly resinous juniper tending to pine with hints of coriander. Soft on the palate with an immediate typical note. Vivid and herbaceous, finishing with characteristic hints of sloe gin. A rich red and intense drupe, with a long resinous finish.

Valbonë, Albania:

Juniper and soft crupe on the nose. On the palate it evolves abruptly, with hints of sweetened dark cherries and drupes. The juniper presents in the background, slightly astringent with a long, resinous finish.

ANGELICA
Angelica Archangelica

Distribution: Northern Europe.

Propagation: seeding.

Parts used: roots.

Harvest: angelica roots are harvested in early autumn, while the fruits are gathered from the umbels in summer. The stem is harvested in late summer and the leaves at the end of spring.

Proprieties: antispasmodic, calming, carminative, digestive, tonic, expectorant and anti-inflammatory.

Curiosity: according to legend, the archangel Raphael brought angelica to earth so humans could enjoy its miraculous virtues. In Hebrew, Raphael means "God's medicine", and therefore the angelica, later "Archangelic", can be rightly called a "divine medicine".

Given its similarity to other umbellifers like poisonous hemlock (which gives off an unpleasant urine smell and whose leaves are very similar to parsley), a certain amount of attention must be paid when gathering it.

Notes: bitter and pungent combination of juniper and celery, as well as lending a "dryness" that helps other ingredients to bind.

IRIS OR ORRIS ROOT
Iris Pallida, Iris Germanica or Iris Fiorentina

Distribution: Eastern and Southern Europe (Italy above all), Morocco and even Asia.

Propagation: shoots.

Parts used: roots.

Harvest: carried out 3 years after planting, between mid-July and mid-September.

Proprieties: antiseptic, flavoring, expectorant, refreshing.

Curiosity: the name derives from the Greek word iris, meaning "rainbow".

One part of the cleaned rhizomes is peeled entirely by hand and using a small pointed, twisted knife.

The iris obtained is sun-dried on cane mats or nets. It is then stored and put on the market after around 5 months.

Notes: woody raspberry/violette. It is an important fixative/binder of other botanicals.

LOTUS FLOWER
Nelumba Nucifera

Distribution: Asia and Australia.

Propagation: water plant.

Parts used: flowers, seeds, leaves.

Harvest: June/August.

Proprieties: the plant, almost entirely edible, is used in the East as an astringent, stomachic, tonic, fever-reducer, hypotensive and vasodilator. A decoction of its flowers is used to treat stomach cramps. The stem is considered hemostatic and is used to treat gastric ulcers and postpartum bleeding. The decoction of its fruit can be taken for agitation, fever and heart problems, while the seed is a powerful sedative.

Curiosity: the flower is sacred in Hinduism and Buddhism, and it is one of the national symbols of India and Vietnam. For Buddhists, the lotus symbolizes the highest level of consciousness, enlightenment, rebirth, and the Buddha. In Hinduism, it represents the center of the universe, and the Hindu goddess Lakshmi, goddess of luck, beauty and prosperity, is said to have been born from a lotus that blossomed on the forehead of the god Vishnu. For the Egyptians, the sun god Ra first appeared rising from the very petals of a lotus flower.

Notes: its most important aromatic components are caryophyllene, pentadecane and methoxybenzene 1.4, which provides the sweet, slightly medicinal scent typical of lotus's well-known fragrance. In gin, it lends splendid warm floral notes.

Chapter IV

The art of distillation

"Civilization begins with distillation"
William Faulkner

We are all familiar with an alembic still. Broadly speaking, its shape and use are recognizable to us, and ultimately we tend to view this instrument and its functions as rather simple. We, the authors, would like to challenge this notion from the offset: distillation, the primary action carried out by the alembic, is not merely a mechanical operation along the lines of "now I add the fermentation, light the still, and heat to 78,15 °C... and now I finally have my ethyl alcohol". Rather, distilling is a science, one that embodies sacrifice, dedication and field experience, and is influenced by countless unimaginable variables. Thus, our first and foremost objective is to explain the many different challenges faced by the master herbalist and the master distiller, as they strive to provide consistency in their products while adhering to high standards.

As any barman will tell you, historically there have been two types of alembic still: the **pot still**, also known as a "Charentais" still, "Répasse" or "double still", and the **column** (or **patent**) **still**, also called a "continuous distillation still", "Coffey still" or "column still". It is in the alembic (a term derived from the Arabic *Al-Ambiq*, which literally means "conical vase") that the "magic" happens: that is,

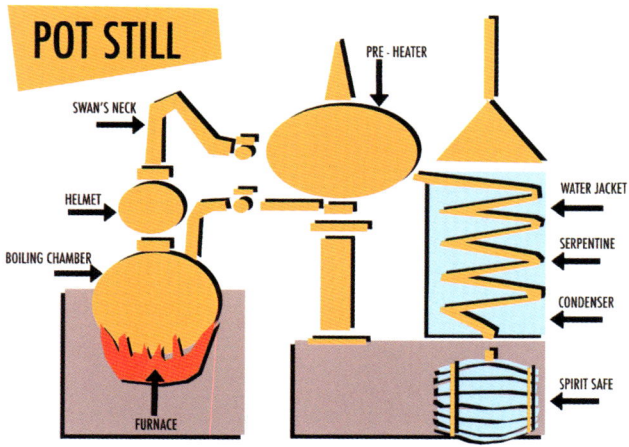

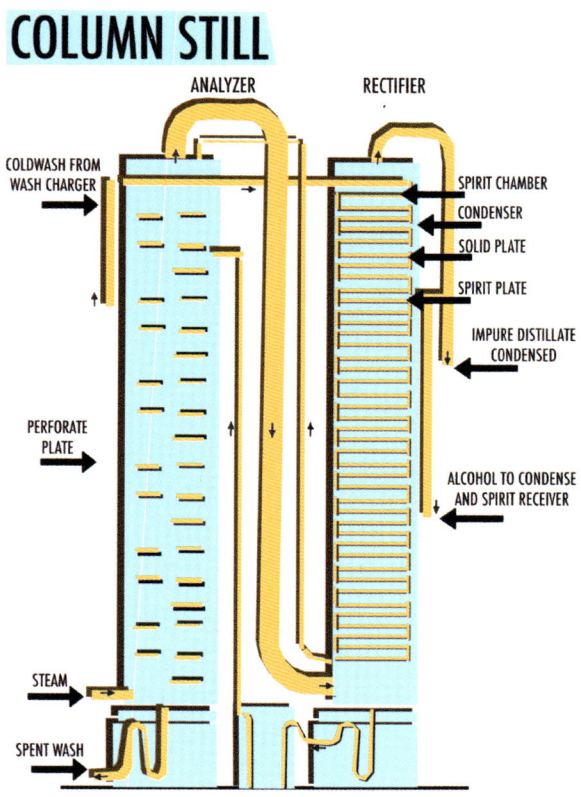

the transformation of a liquid into a gas, and then again into a liquid. This transformation is brought about with heat; the alembic is essentially heated up. The compound contained within is fermented, a product that already contains ethyl alcohol and an infinite array of other precious and non-precious alcohols that comprise it. Fascinatingly, each of these alcohols can evaporate, divide or separate from the compound at different temperatures. By virtue of this, we can not only obtain different volatile substances but also select them and separate them... this, in short, is what constitutes the greatness of the alembic still.

Now, it's important to understand that from two identical alembics, made by the same company, using the same materials and being of the same size, two different distillates will emerge. This is because the still is to the master distiller what the paint brush is to the painter, something like the orchestra conductor's baton. Meanwhile, the human factor is always the determining factor in distilling. Moreover, stills are not standardized; each distillery studies and plans its own, based on its specific needs. In other words, you could say that each and every alembic still is "tailored".

Dispelling a commonly held though not entirely incorrect belief is vital to this book's primary focus—gin. Differently from whisky, cognac, tequila and mezcal, the process of making gin does not start with distillation, but rather with re-distillation. At least, this is true in some 95% of cases. Recall the accurate definition of gin as "a compound distillate produced using neutral food alcohol, to which aromatic compounds called *botanicals* are added". Distilleries that produce gin purchase a base distillate, which often has been distilled twice (those aiming for a premium, super or ultra will acquire a base that's been distilled three or even four times) and then re-distill it with the added botanical component.

In the absence of requiring or seeking particular notes or aromas drawn from the raw material of the fermented product (as is the case for most gins, but not all, mind), distilleries simply choose a distillate to use as the base. Otherwise, those distillers who do produce a base themselves will drastically lower production costs.

In most cases, what we are talking about is a grain alcohol base, typically made with corn, wheat and barley. Then, little by little, oats, rye, spelt or sorghum are included. These starter distillates are rarely 100% mono-grain, but rather a blend that will form the product's supporting structure. An example of this blend could be: 85% corn, 10% rye, 5% barley.

In this world, it is the parent company that invariably decides the procedure, while considering the quality and type of alcohol they wish to obtain for their product. Keep in mind the global market features gins whose alcohol base is not grain-derived but instead derives from other fermentable organic substances, like grapes, cane sugar/molasses, beets, bananas, apples, cow's milk, and so on...

Given the starter distillate, the question now becomes one of understanding how this base can be transformed into a gin. Not surprisingly, there is no one way to do this. But before delving into techniques, a few important premises should be outlined.

This first of these, worth mentioning again, is the synergy between the qualitative research carried out by the master herbalist and the technical-extraction skills of the master distiller, all the while with these parameters guiding them:

1) **the alcohol content:** alcohol acts as an extractor, so the higher the alcohol content, the greater the extraction. Greater, but also faster, meaning a factor to consider in relation to the botanicals as they macerate. Flowers require a different alcohol content than rhizomes or bark, for example, which is hopefully a helpful specific example of the concept.

2) **the temperature:** the degrees centigrade of the compound influences the extraction prior to distillation, a phenomenon that can be observed while preparing a hot herbal tea and a cold one.

3) **the amount of botanicals used:** botanicals can vary in both quantity and quality, depending on the harvest time, climate conditions, type of transport, duration and storage prior to being used. Since it is practically impossible to acquire the same quantity and quality of botanicals each year, management of this aspect of production further highlights the master herbalist's knowledge and skill.

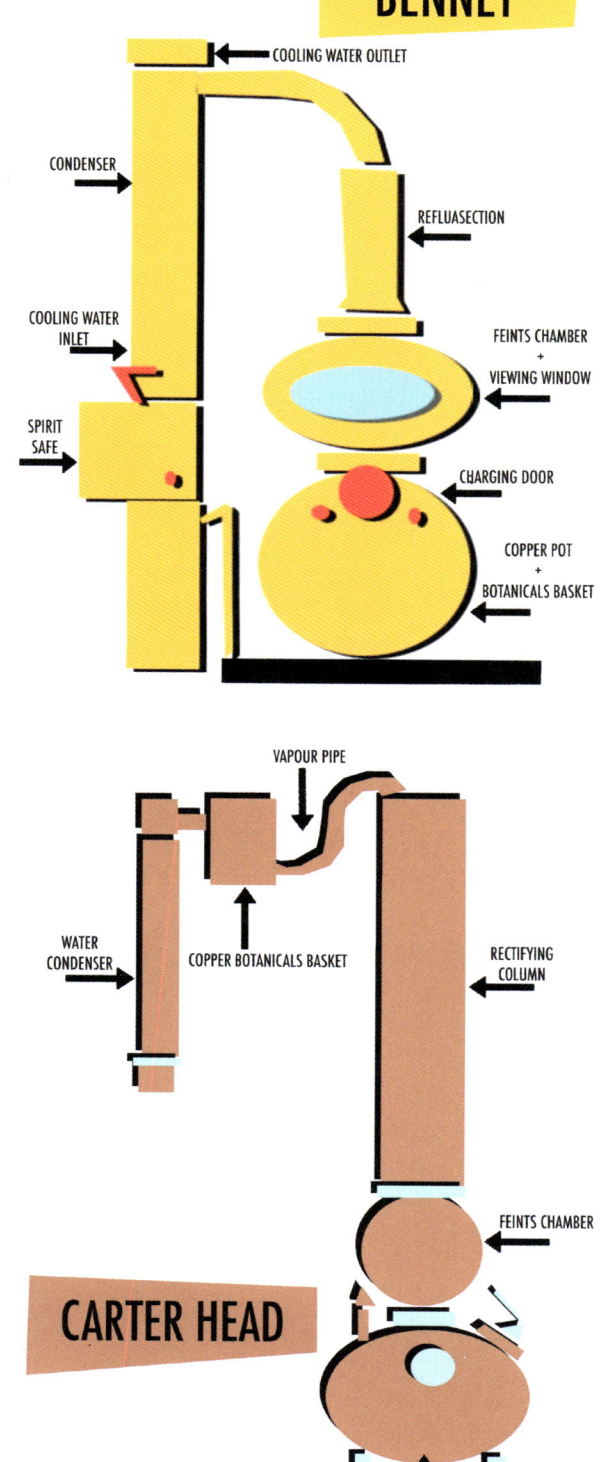

4) **the extraction technique:** many techniques exist to extract aromatic components from botanicals: infusion, decoction, percolation, maceration, and of course, distillation. Each technique has different capacities when it comes to extracting the gustative and olfactive features of each botanical.

Such a short list should suffice to help us grasp just how complicated the production of quality gin really is. With almost infinite variables to manage, and the annual onset of new challenges and difficulties to face, producing a quality gin involves significantly more than merely choosing an alcoholic base and a bouquet of botanicals (here, too, the temptation for some is to just exaggerate the amounts, with the idea that more equals better). Such simplified notions of gin are trivial indeed, as well as decidedly misleading. For comparison, let us take an example from the domestic realm: making a cup of tea. What temperature should the water be? How long should the tea steep? What type of tea is being used? While perhaps few of us ask these questions during the practical act of making tea, they are questions that reflect an implicit desire to work with a tea's specific properties and prepare the best cup possible. Let us now analyze the distillation process in the same way, using a classic structure gin like a London Dry as an example, to consider three different distillation techniques. The first, called "steeping", involves immersing the botanicals directly in the base distillate in the alembic boiler, using a pot still, a **Bennett still** or a generic copper still. So, how and when does one place the botanicals in the base while using this method?

Botanicals can be immersed all together randomly, or collected in a canvas, gauze or jute sack and left to soak. Note that they can also be placed in the hydro-alcoholic solution at different times. If, for example, a recipe calls for 40% bark, rinds and rhizomes, then 40% seeds and fruits and finally 20% flowers and leaves, you could add the tougher items first. Then, after a period of time that close study and practice will dictate, add the seeds and fruits. Lastly, at an even later stage, more delicate botanicals such as flowers and leaves can be added. Of course, all of this takes place prior to distillation. The temperature of the base alcohol, the required alcohol content, and the amount of time to leave the botanicals in the alcohol are all decisions each company and every master distiller must make—all decisions that will influence the final product. Some choose to leave botanicals immersed for just a few hours, while others might "forget about" them for as long as a few days.

Next is the technique known as "raking". Less invasive than steeping, raking consists of placing all the botanicals in a dedicated basket or grid-like component. This in turn is hung inside the belly of the still, without ever coming into contact with the hydroalcoholic solution; it hangs slightly above the liquid. Here, extraction happens in a more delicate, less invasive way compared to techniques that involve contact with the liquid. With the alembic activated, the hydroalcoholic solution boils and the steam passes through the holes in the basket, passing

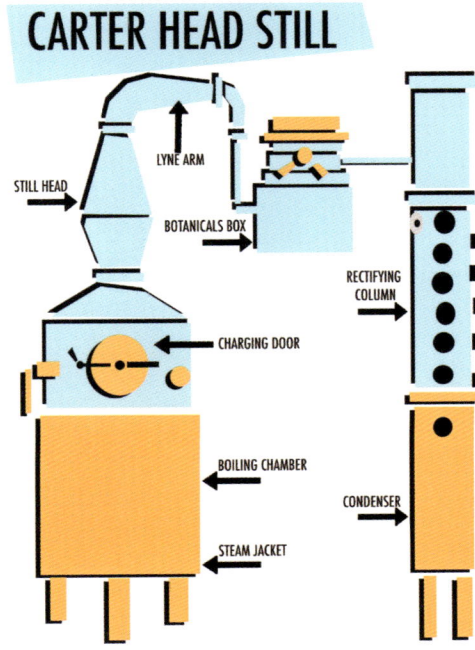

through the plants themselves. This captures their various scents, which are then returned to a liquid state in the re-condensation phase. All the vapors (heads, hearts and tails) then pass through this botanical cocktail and are recondensed and selected, extrapolating the heart, which is subsequently lowered by degrees, stabilized and finally bottled. Incidentally, there are stills capable of combining both these techniques, and these allow the master distiller to both immerse some botanicals and place others in the basket or on the grid.

The third technique, even less invasive and more delicate than raking, takes its name from the still used when following the technique itself. This is the **Carter Head Still** (p. 101), a type of still developed at the start of the 19th century by the Carter brothers,

AENEAS COFFEY AND THE CONTINUOUS DISTILLATION REVOLUTION

The Industrial Revolution was a period of enormous scientific and technological progress. With regard to the world of distillation, this period saw the arrival of one of the most important inventions within the sector, the continuous distillation still.

Several individuals from different countries were involved in the development of the still that would significantly contribute to improving the quality of the gins on the market: Scottish Robert Stein, Italian Bartolomeo Baglioni, Dutch Jean-Baptiste Cellier Blumenthal and French Augustine-Pierre Dubrunfaut. Each of these individuals claimed to have invented the continuous distillation still, but history and right have granted this honor to only one person, the French-Irish man named Aeneas Coffey. The column still bears his name even today, if for no other reason than the patent he officially filed on its behalf, with the registration code 5974, in 1830.

In reality, the other figures mentioned here were legitimate participants in this story, hardly liars or braggarts! In 1813, Cellier Blumenthal had registered the first patent for a multiple column still that promised to supplant the classic slow and laborious pot still. His studies paved the way for Anthony Perrier, an Irish distiller from Cork who took this concept and applied it to the production of whiskey. In his column still, the mash was added gradually, in small quantities. It then passed through a very narrow path, allowing for much higher temperatures and thus obtaining a greater selection of higher alcohols. Perrier inspired Robert Stein in turn, who in 1828 created the patent still, which could feed the "wash" directly to the distribution column. This system was tested for the first time at the CameronBridge Grain Distillery in Fife, Scotland. Unfortunately, Stein was never able to obtain the funding needed to take his patent outside the area, even though, according to some customs inspectors of that time, his machine turned out to be the best of its kind, in terms of the final product created. Dubliner Aeneas Coffey had a particular advantage on his side. Working as a customs inspector, Coffey was in a perfect position to observe the various related projects and patents being developed during his time. He noted their strengths and related problems, before finally making the big leap: the design and construction of his first alembic still for the Kilbeggan Distillery. Later Coffey opened the Dock

Distillery, located on central Dublin's Grand Canal Street. His establishment boasted two distillation columns that improved the liquid separation system, eliminating the problem of re-distillation and ultimately obtaining a noticeably cleaner, lighter product, in addition to its higher alcohol content.

Coffey's still was so successful, very soon all producers wanted to work with his model. Just five years after he'd registered his patent, Coffey closed his distillery and went on to found a company dedicated exclusively to the manufacture of stills in London. Today, Aeneas Coffey & Sons is still in operation, under the well-known name of John Dore & Co. Limited. Oddly enough, the international success of Aeneas Coffey's continuous distillation still was not matched in his homeland: Ireland remained faithful to the triple distillation method, as it still is today. But the Coffey still was revolutionary. Remember, before this achievement, adding sugar to distillates was the customary way to make them drinkable, such as with Old Tom. The Coffey still changed that, lessening the need for added sugars and paving the way for the current great protagonist in the world of juniper-based spirits: London Dry Gin.

whose precise purpose was the making of gin. Having studied under the noted Aeneas Coffey, inventor of the Patent Still, from whom they learned the craft, the Carters decided to strike out on their own. Legends about this still abound, most of which are meant to prop up this still's general reputation. One that interests us particularly relates to understanding the still's functionality. Similar to a Bennett still, the Carter Head still has a slightly longer separating column. But the true differentiating feature of this unique still is the position of its basket (or botanicals chamber), which is located outside the fractionating column rather than inside, just near the recondensation column. Such a design permits only the noble (or selected) vapors to pass through the different chamber cells, inside of which the botanicals are enclosed (for the most part, these are inserted in studied weights-per-gram amounts, but not in a precise sequence, in order to attain an almost homogeneous extraction).

This historic extraction method is still used and copied by other developers to obtain certain extractive notes. It is also known by the term "vapor infusion", which obviously is not synonymous with the **Carter Head** (p. 100), but rather a reference to how the botanicals are collected in the dedicated chamber—outside the body of the still—through which the steam produced during distillation passes.

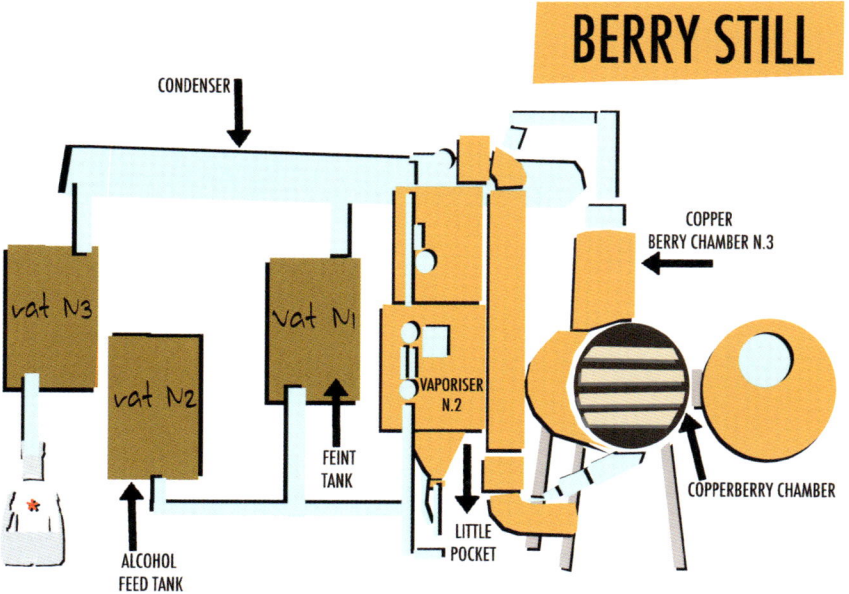

Lastly, yet another type of alembic merits mentioning, as this one uses both systems. The Berry Tray dates to the 1920s and was originally designed for use in the production of perfumes. Today, a famous example is found at the Balmenach Distillery, where it is used in the making of their noted Caorunn Gin.

The Berry Tray's contact chamber between the botanicals, and the steam is located at the base, a rather non-traditional placement. It is cylindrical in shape, with an internal diameter of some 91 cm, and positioned horizontally. Inside are four perforated trays in stainless steel, inside of which the 20 kg of botanicals needed for this particular recipe are placed. During distillation, the trays are carefully moved closer to the steam being produced, extracting the aromatic components.

The procedure is as follows: this is a small-batch alembic, known in some cases as "artisanal" or "craft still", with dimensions that allow for manual management, something that with stills of other sizes is much more difficult. Its precise capacity is equal to 1,000 liters of neutral grain spirit stored in a dedicated vat labelled No. 2. The alcohol is channeled to the column still, also called "Vaporiser No. 2" (and commonly referred to as "Little Rocket"), where the heat transforms the alcohol to vapor, in turn channelled to the base of the Berry Chamber. Contact takes place within this chamber, resulting in the extraction of botanical aromas. Everything is then channelled towards the upper part of the Berry Tray and directed towards the condenser, which re-condenses the distillate and gathers it in Vat No. 1. The process continues until Vat No. 2 is completely emptied. At this point, without turning off the still, the master distiller empties the entire contents of Vat No. 1 into Vat No. 2, to repeat the distillation/aromatizing

Chapter IV

a second time. In this way, the distiller is able to extract every possible aromatic component and create a "concentrated botanical flavor" which, when at last it is re-condensed again, is stored in Vat No. 3. And it is now that it takes on the name "concentrated Caorunn Gin".

From the original 1,000 liters, around 945 liters of full alcohol content gin spirit is obtained. Now the distillate is chemically analyzed and, once authorization is acquired, the degrees are lowered to 41.8% volume. Then it is bottled.

After reading descriptions of these many and various techniques and alembics, one might wonder why all these differences exist. After all, we are talking about making gin here, not something akin to painting the Sistine Chapel, right? The reason, however, is for me very easy to understand.

Sometimes, the "on paper" idea of something does not reflect the best method for achieving excellent results. Hundreds of considerations must be taken into account, from the procurement of raw materials to the distillation volume, then the production locale, the type of alcohol used, the effects of differing costs, the sale price established for a given market, and many others. Above all though, a "mother" question stands out as the one every master distiller must ask before launching such a complex and demanding endeavor. It is a question upon whose response the various elements described here must depend: "What do I wish to obtain?". It's a valid question, because ultimately every gin must absolutely be matched with its associated identity, based on an infinite array of factors—factors that can improve or compromise that identity. Some excellent products are available on the market that are made from a basic double-distillation alcohol, for example. If these had instead been made from a four-distillation alcohol, decidedly superior "on paper", they would still not have acquired their distinctive and wonderful notes, their identity. To take the point further, think of how common it is today to associate a superior product quality with a high quantity of spices used, and how we attribute better quality, almost instinctively, upon hearing or reading: "This gin contains 55 botanicals!". Somehow, in our minds a certain product is immediately superior in some way, even though the amount of spices used in a gin recipe rarely means better quality or greater complexity.

Some gins possessing rich aromatic complexity are produced using only three botanicals, an example that helps us appreciate the masterful work and requisite synergy between master herbalist and master distiller: that is, between the person who handles the selection of botanicals, and the person who determines and guides the distilling process. The rest is marketing.

With these considerations in mind, we wish to highlight a small yet not unproductive controversy, one related to the new tendencies taken up by some barmen: adding an undefined quantity of various botanicals to a Gin and Tonic. This habit results merely in compromising the hard work, experience, and choices made by the two masters.

Remember, a Gin and Tonic's balance can be ruined simply by adding the wrong amount of ice!

Other slightly more modern distillation techniques are worth discussing, the first of which being individual botanical distillation. This technique is used primarily in the distilled compound style. Undoubtedly the simplest type of distillation, it acts upon the individual botanical, making the most of its characteristics, rather than distilling all the botanicals together. It is a system that provides decidedly greater continuity in a recipe. For example, with a gin that will be composed of eight different botanicals, the needed quantities of each botanical distillate are prepared (botanicals gathered at their peak aromatic moment) and stored separately in steel tanks. When market requests for volumes arrive, one simply taps each tank containing the required distillate, creates the compound (the recipe blend, in other words), and bottles it. So what is the added value of this technique? With this method, the final product is greatly influenced by the raw materials having been harvested while at their peak period, thus obtaining a practically perfect continuity and balance. The best known example among gin enthusiasts is Gin Mare, a gin that is the result of a perfect union of eight different distillates and a likewise number of individual raw materials. Produced by Vantguard, this gin's most distinguishing ingredient is the Arbequina olive, harvested at the best aromatic moment possible of course, washed well and let to macerate directly in alcohol for more than 24 hours. The alembic is lit, and the obtained exquisite olive distillate will then be added to the remaining recipe ingredients in precise percentages. This system allows for monitoring every single maceration in the alcohol and every single distillation, with the best possible results. Naturally, other businesses also rely on this system: Rivo, Windspiel, Death Door, Bobbys, Luz and Gin Primo, to name a few.

Completely different from individual botanical distillation, the technique known as "multi-shot" is used principally by companies producing gin in very large quantities. Here, quality is maintained both in the base alcohol and in the botanical component, yet the goal is not so much to create a "hand-crafted" or "small-batch" gin but rather to produce large volumes, thereby guaranteeing a more competitive market price. Whereas the one-shot method is the simple realization of a recipe, appropriately distilled with the exact quantity of each botanical, the multi-shot is precisely the opposite: using the same amount of alcohol, the proportion of the botanicals is multiplied to create a super-concentrate. This will in turn be combined with neutral alcohol, up to the concentration required by the original recipe. Essentially, this method means greater efficiency, considerable increase in production, moderate energy savings and a reduction in production costs—all factors contributing to a better shelf price. Many excellent international brands of gin are produced following this technique.

In the following pages, we turn to our friend and colleague Giovanni Ceccarelli, who will outline two additional "modern" techniques currently attracting great attention from operators: vacuum and cold distillation. Moreover, we wish to conclude this chapter dedicated to distillation by attempting to solve a few legitimate curiosities. The first relates to copper, the material most commonly used in the making of gin alembics, first and foremost on account of its natural aptitude for conducting heat. Of course, some will argue that gold is even better than copper here; at the same time, one could counter this by pointing out the sheer folly of using such a fine and costly metal to make an alembic still. Furthermore, copper is a ductile metal, one easy to work into any shape—a detail of no small importance to those who build these devices. Copper is also highly resistant to corrosion, and thus does not leak foreign flavors into the distillate (on the contrary, copper acts as a catalyst in the promotion of esters, which lend fruity notes to the final product). Finally, copper is undoubtedly the best material for an alembic used for aromatic distillates, given its ability to affix foul-tasting acids produced during fermentation and form insoluble salts called "copper sulphate" or "copper salt", which are in turn deleted during the process.

Often an alembic bears its own special name. Some of these are outlined in this current chapter (such as Little Rocket of Caorunn) and in a later chapter dedicated to the 100 not-to-miss gins. These include: Angela, Jenny and Constance of the Langley Distillery, Prudence and Patience of the London's Sipsmith, Christina and Little Albion at the London Distillery Company. This typical custom among companies is a testament to (if one were still needed) the intimately close and even affectionate relationship that binds the master distiller and his team to the alembics with which they work daily—in other words, a drop of poetry within the magical world of distillation.

"Anomalous distillations": Vacuum and Cold

In collaboration with Giovanni Ceccarelli, Drink Factory trainer, bartender and consultant

VACUUM DISTILLATION

Reduced pressure distillation, more commonly known as "vacuum distillation", is a delicate topic, and one not easy to immediately comprehend. I will, therefore, try my best to simplify the concepts as much as possible.

To begin, though the instrument in question involves distillation, it really has no direct relation to the traditional alembics we are all familiar with. This is not a still made of steel or copper. Rather, its "raw material" is glass, and it is more like a laboratory instrument than a "still room" instrument, although some large industrial versions do exist today. To better clarify the concept, it's worth referring to the rotary evaporator or **Rotavapor** (p. 111) invented by Walter Büchi in 1957 and today a registered trademark. Obviously other businesses are active in this market, such as IKA or Heidolph. The uniqueness of this instrument derives from its design: it is crafted to be connected to a vacuum pump which, by sucking air from inside the distillation circuit, reduces the pressure within the system, and consequently lowers the boiling temperature (evaporation) of the liquid treated.

To further elaborate, let's consider the variations in temperature when boiling water: at sea level, water boils at 100 °C, while in the mountains, for example at 4,000 meters above sea level, it boils at 86 °C. This happens because in the high mountains, the atmospheric pressure is less.

The extraordinary thing about the rotary evaporator is that it allows for distillation to happen even at temperatures below 20 °C. The many advantages of such a unique feature are clear. Firstly, the energy savings, given that the temperature does not have to exceed more than 78 °C, as it does with common stills. Naturally, if used in a reduced capacity, such as in a bar or a home kitchen, the energy savings will likely be somewhat negligible. Yet there are other significant aspects to note.

Reducing boiling temperature allows for distillation at a low temperature, and distillation at a low temperature in turn allows you to preserve all those substances sensitive to this parameter. Then, in addition to lowering the boiling point, another aspect of great importance is the minimal amount of oxygen inside the circuit. In this way, the wasteful processes caused by the presence of oxygen are significantly reduced.

To summarize, vacuum distillation involves a greater respect for those substances that can be altered by the presence of oxygen or by high temperatures.

And it is due precisely to its low-temp operation that this tool can be used to make cold reductions, such as fruit or vegetable reductions (when water is removed from fruit juice, it becomes concentrated and thus much more flavorsome). Think of a Daiquiri made with concentrated lime juice, a Satan's Whiskers made with concentrated orange juice, or a Bloody Mary prepared with fresh, then concentrated, tomato extract (obviously we are not talking tomato paste, such as the kind sold in tubes). Then, with this form of distillation, it is possible to distill colors obtained from spices or aromatic herbs. For example, after extracting mint in alcohol, a transparent, mint-flavored distillate can be obtained, but one that does not have the typical bitter aftertaste of this particular herb.

Ian Hart uses this type of instrument to produce his Sacred Gin. Hart distills each individual botanical separately with a pressure level that varies from $1/12$ to $1/6$ from the atmospheric pressure, depending on what he wishes to extract and the type of botanical being used. Distilling by this process and at this pressure means the various aromatic notes and freshness are not compromised by the heat's action. With regard to temperature, Hart distills between 25 and 50 °C, with a variation dictated by the botanicals. Based on his experience, 40 °C is the highest extraction temperature to use with botanicals, to avoid damaging the aromatic component—a fundamental aspect to bear in mind.

COLD DISTILLATION

Extreme cases of low pressure distillation can lead to what is erroneously called "cold distillation". One of my favorite distillers, Oxley, has used this technique to produce his very own masterpiece. Here, a different tool is used from the one used to make Sacred Gin, yet this one still allows for reducing the pressure inside the evaporation circuit. By drastically lowering the pressure, Oxley is able to distill at a good −5 ° C, which is precisely why they call it "cold distillation"; however, the physical process is the same, and the instrument's operating principle is entirely identical to that of the rotary evaporator.

While the evaporator used for Sacred can bring the pressure to 40–50 Torr, the one used by Oxley is able to arrive at 7–9 Torr. To provide some perspective, atmospheric pressure is equal to 760 Torr.

To conclude, and in the interest of being completely sincere and transparent, I will say this: I believe it is appropriate to remember that lower temperatures do not always guarantee better results when it comes to flavor. As is true with almost all crafts, in our profession it is knowledge and experience that guide us in making consistently accurate selections, and therefore obtaining the absolute best characteristics from each of these selections.

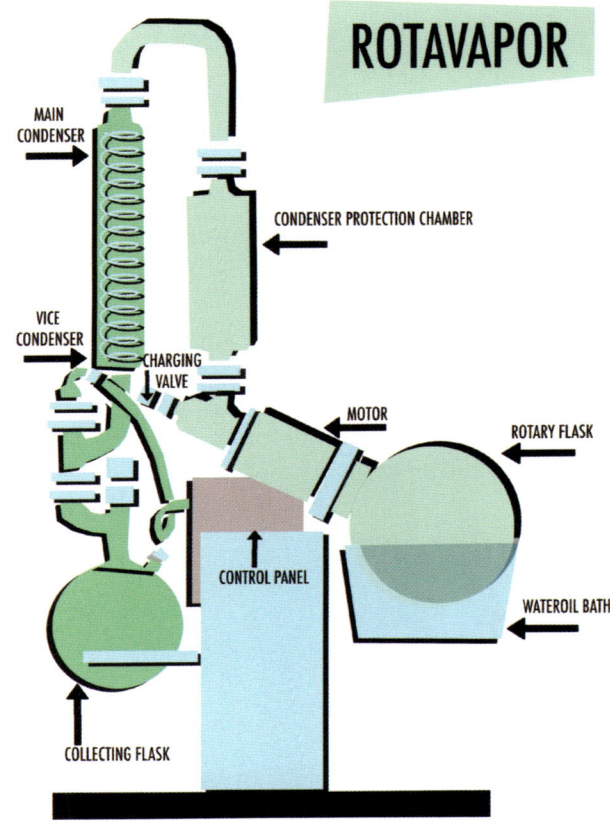

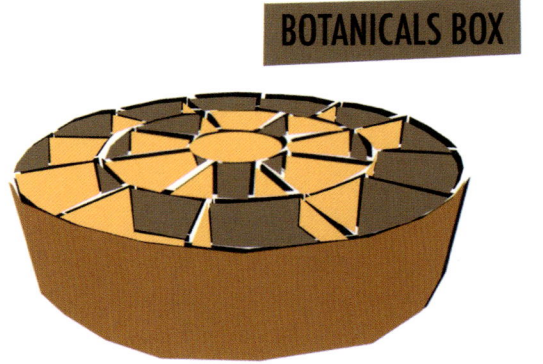

GAY-LUSSAC, AMERICAN PROOF AND SIKES

Checking the alcohol content on a bottle is always a good idea. Typically, the information presented here is fairly easy to understand. Although, perhaps not everyone is aware that more than one alcoholic strength scale exists, a very real factor that can present some risk and potential for confusion. And given that these are alcohol percentages, being absolutely clear on this matter is vital. In Italy and numerous other countries throughout the world, the "alcohol scale" bears the name Joseph Gay-Lussac, a French physicist and chemist known for the laws of gas named for him. His standard of measurement indicated, at zero degrees Gay-Lussac (or 0 °GL) was a distillate of pure water, while at one hundred degrees Gay-Lussac (or 100 °GL) was a distillate of pure alcohol, today written as % vol or % abv. Gay-Lussac created correlation tables on behalf of the French government between the relative density (with respect to water) at 15 °C and the composition of hydroalcoholic mixtures expressed as the volumetric concentration of the ethanol present. This novel method introduced by Gay-Lussac, of referring the relative density to volumetric and non-weight ratios, has proved to be the most suitable solution for practical and commercial applications.

In the United States, you will still see products bearing the so-called "American Proof", another scale that can be compared to the Gay-Lussac scale very simply: just divide the American proof in question by two. Example: 100 American Proof is equal to 50% vol or 50% abv. The final existing scale to discuss (also found on the comparative table below) is one that takes its name from the Englishman Bartholomew Sikes, the use of which remained in force in Great Britain until 1980, thus you can still find it on bottles dating to before that year. However, as British producers are not obliged to comply with the newer discipline, knowing how to transform Sikes measurements into Gay-Lussac is a good idea. Here the division is by 1.75. For example, 87.5 Sikes is equal to 50% abv.

Comparative scale of alcoholic strength											
Sikes	0	17,5	35	52,5	70	87,5	105	122,5	140	157,5	175
Gay-Lussac		10	20	30	40	50	60	70	80	90	100
American proof		20	40	60	80	100	120	140	160	180	200
	Water	Table Wines	Fortified Wines		Normal Spirit						Absolute Alcohol

Chapter V

Gin and its rules

"I feel sorry for people that don't drink, because when they wake up in the morning, that is the best they're going to feel all day"

Frank Sinatra

Although arguably the most creative of distilled spirits in the world, even gin—the distillate characterized primarily by juniper—has some rules to consider. And knowing them is important, if for no other reason than to be part of the conversation and understand what's going on behind the bar.

But before delving into this complex world of gin's rules, regulations and matters relating to its classification, let's start by outlining a definitive definition of this distillate.

Gin is "an acquavite or distillate made from grains and/or starchy substances, flavored predominantly with juniper berries along with other spices commonly defined as 'botanicals'". It has a strong, decided character, often pungent, and colorless. Depending on the country of origin, gin is produced by distilling a fermented grain product, wheat and barley, and also grapes (G'wine, for example), apple cider (Chase Gin), sugar beet or sugar cane (Monkey47), potatoes (Windspiel) and more. As part of its production, a blend of herbs, spices, plants and roots is let to macerate in a fermentation, and among the various ingredients, collectively referred as "botanicals", juniper (*Juniperus Communis*) has always been the undisputed star of this prince of the white spirits.

According to the EU policy guideline of 17 April, 2019, the word "gin" indicates all spirit beverages obtained by flavoring ethyl alcohol (of agricultural origin and/or grain brandy and/or grain distillate) with juniper berries (*Juniperus Communis*) that contain a volumetric alcohol percentage equal to 37.5% or higher. At the discretion of the producer, several other aromatizing substances can be used in the botanical mix, and indeed more than one hundred and twenty comprise this group known as "botanicals" (including angelica, coriander seeds, ginger, citrus peels, cinnamon, licorice, cumin, anise, cardamom, cassia, iris, almonds, fennel, nutmeg, orris root, cucumber, roses). Yet, again, in every instance, juniper must be the main botanical.

The primary regulatory guideline on gin is the so-called "Spirit and Drink Regulation", or EU Regulation 2019/787 (available online), whose application obviously falls solely within the European Union alone. Of course, gin is regulated in other parts of the world, too, sometimes with differing rules but often with several overlapping points, as is the case in the Far East and the United States (regulated since 1991 by the U.S. Bureau of Alcohol, Tobacco and Firearms) and Canada (the same, in 1993) and Australia (1987).

Juniper Spirits

Juniper spirits are those distilled beverages obtained by flavoring ethyl alcohol of agricultural origin and/or grain spirit and/or grain distillate with juniper berries (*Juniperus communis L.* and/or *Juniperus oxicedris L.*).

The minimum alcoholic strength for this type of product is 30% vol. Other natural flavoring substances, and/or substances identical to natural ones, and/or aromatic preparations, and/or aromatic plants (or parts of them) can be used to complement these drinks. Whatever the amounts, juniper must be perceptible among the organoleptic characteristics, though it is acceptable if the juniper is perceivable to a lesser degree. Juniper spirits can be sold under the names "Wacholder" or "genebra".

Gin

Gin is the juniper spirit obtained by flavoring ethyl alcohol with juniper berries (*Juniperus communis L.*). The alcohol must be from agricultural origins and have the appropriate organoleptic properties. The minimum alcohol content of gins is 37.5% vol. Gin production requires that only natural flavoring substances and/or substances identical to natural ones and/or aromatic preparations can be used. Here, again, juniper must be the predominant flavor.

Distilled Gin

Distilled gin can be:
1) the juniper spirit beverage obtained exclusively through the redistillation of ethyl alcohol (of agricultural origin of adequate quality), possessing the appropriate organoleptic characteristics, and with an initial alcoholic strength of at least 96% vol, using traditional gin alembic stills, and with juniper berries

present (*Juniperus communis L.*) along with other plant products, provided that juniper is the predominant flavor.
2) the mixture of products from said ethyl alcohol distillation of agricultural origin and like composition, purity and alcoholic strength. To flavor distilled gin, natural flavoring substances and/or substances identical to natural ones and/or aromatic preparations can also be used. The minimum alcohol content of distilled gin is 37.5% vol. Gin obtained solely by adding essences or aromas to ethyl alcohol of agricultural origin is not distilled gin.

London Gin

London Gin is a type of distilled gin obtained exclusively from ethyl alcohol of agricultural origin, whose aroma is due exclusively to the redistillation of ethyl alcohol in traditional alembic stills, with the presence of all the natural plant materials used. Its alcohol by volume content must be equal to or higher than 70% vol. It may not contain sweeteners in quantities higher than 0.1 gr/l in the final product, nor colorings or other added ingredients aside from water. The minimum alcohol content of London Gin is 37.5% vol. "Dry" is added to the name, becoming "London Dry Gin". The following summary should help further facilitate an understanding of classification and how it applies to our beloved gins.

London Dry Gin (L.D.G.)

Also known as "London Gin", the name of this gin has always carried some confusion with it, primarily on account of the word "London". The term actually refers to a style; London Gin is not made exclusively in the English capital city. It can be produced everywhere.

What makes this complex and historical gin style one of the most unique in the world comes down fundamentally to one point: namely that "all aromas (or botanicals) used must be natural (in other words, no oils, essences or tinctures) and these must be transmitted during the distillation process". It is this point and this point alone that makes London Dry Gin one of the world's most complicated gins to produce—it must be able to balance the full aromatic profile resulting from the extraction of the chosen botanicals, which will obviously differ in aromatic strength based on their individual characteristics. One advantage producers of London Dry Gin have in their favor is the choice whether to work with the one-shot or the multi-shot method (see chapter 4 on distillation), thereby significantly reducing production costs. To slightly "dilute" the aromatic charge, small amounts of 96% abv neutral alcohol base, also known in common jargon as NGS (Neutral Grain Spirit), of the same quality (if not better) can be added to that used for the production, and water can be added to lessen the alcohol content. However, adding any other type of aroma is absolutely prohibited, whether fresh or artificial. This gin may not be colored in any way, but sugar may be legally added in amounts equal to 0.1% gr/l. Since we are here, let us also clarify a point that I believe has caused great confusion: Can you blend two different London Dry Gins? Perhaps even obtained from different botanicals, as long as they are natural and the dominant component is still juniper? The answer is yes, of course. Some London Dry Gins are blends from two different alembic stills, stills often of different shapes or even ages, and made from differing botanical components that are then combined. A classic example of this is Bickens Gin, the result of two different distillate recipes made in two different historic alembic stills, each with completely different capacities.

Distilled Gin

Distilled gin is also known as "distilled compounds" or, to cite a wonderful definition provided by Ryan Magarian in April of 2009, "New Western Dry Gin", which later became "Contemporary Dry Gin" or "New Wave" or "New American Dry Gin". Such definitions clearly demonstrate to what extent gin has been shedding its former more obsolete conceptualizations, thanks to new techniques, investments and creativity that have entirely overhauled this "king of white spirits".

Diving into more specific details of this discipline, one aspect that immediately catches the eye is the wondrous array of possibilities when it comes to selecting aromas. This is likely also one of the discipline's more delicate points. To begin, a clear distinction should be made about aromas, whether they are essences, tinctures or oils, whether high quality or not. This topic alone could be enough to write significantly about. Then, their various applications should be evaluated, given that one of the prerogatives of this category is that the product be redistilled. Allow me to better explain: gins produced from a double-distillation base alcohol with the addition of mixed aromas absolutely cannot be included in this category (this can be done of course, only not with distilled compounds). Having clarified this point, however, we must also remember the saying that "all laws are meant to be broken". In other words, one could actually redistill the neutral alcohol with just a few juniper berries of questionable quality and origin, and then, after distillation is complete, add your own aromas. With this method, one could produce a wonderful (so to speak) distilled compound, but at a price. Fortunately, however, some companies operate with total respect for the consumer while adhering to foundational principles of quality and professionalism.

Working with this type of gin means paying close attention, as we research and work towards maximizing and enhancing each individual botanical's organoleptic qualities. Here, too, allow me to explain better: with this method, one can distill each ingredient separately to maximize its freshness and quality, creating the most precise, qualitative distillate possible, and then mixing this with the other individual distillates (whether single- or multi-botanical).

Some brilliant examples include 2punto4, the new entry from Mauro Mahjoub and Jorge Alberto Soratti, the well-known Gin

Mare, exactly like By The Dutch, Rivo, Primo, Sacred, Bobbys, Windspiel, Leopold's, Forest and many others... all are examples of producers who have chosen to work with individual botanicals, distilling them one by one and then crafting the definitive batch. Consider Ondina Gin, which features one of nature's most delicate botanicals—fresh Ligurian basil—distilled separately and is subsequently combined with a proper London Dry Gin made up of eighteen other botanicals. We could also mention Jinzu, which combines a juniper distillate with a sake distillate, adding only the finest, highest quality essences. Then we have Kinobi Canaïma, a gin for which the botanicals used are subdivided into groups based on their structure and then distilled.

Working with distilled gins allows for some fun to be had, while creating products of the highest quality. Redistillation (rather than a simple maceration process) is essential with these gins that can be colored, unlike London Dry Gins, and sweetened up to a maximum of 0.5 g/l. Lastly, ethyl alcohol of the same origin can be added to them after distillation. The same goes for simple water.

Cold Compound Gin

None of the regulations regarding this category of gin requires redistillation, leaving room for some rather simplifying production techniques, summed up as: add aromas to an NGS, mix well, filter as needed, and bottle. To be fair, however, it should be said that several companies have successfully experimented within the elastic nature of this gin category, creating some true Old Style Gin masterpieces. Old Style, because time ago—when poverty was rampant and the alchemical skills and techniques required to establish a distillate's quality and moreover adequate technology to produce the well-known London Dry Gin had not yet come into existence—this was the undisputed production system, that which involves direct contact between distillate and botanicals. This remains the method followed by many major brands: to macerate the individual botanicals in different batches, naturally at different temperatures and alcoholic concentrations, making sure to filter the mash (as desired) and combine it in the various percentages appropriate to craft the recipe. This method is employed by brands such as Gin del Professore, Roby Marton, Bath Tube and many others who have bet everything on processing alcohol directly with botanicals—a kind of pure magic, with elements of spontaneous folly. As mentioned prior, there is no need to redistill in these cases. Yet, the alcohol base must be of agricultural origin, and at alcohol concentration not exceeding 96% ABV. The aromas used, whether natural or artificial, must always be approved by the Ministry of Health and by the community standards established in collaboration with the European Food Safety Authority (EFSA). Finally, there are no restrictions on the addition of approved additives such as sweeteners and/or colorings when producing this type of gin.

Other Gins

Here we will outline some products that could be considered gin or something very similar to it, including some PGI (Protected Geographical Indication) status gins, or the so-called "non gins".

Mahon or Xoringuer Gin: essentially a true gin, but also a PGI, this gin is made exclusively on the Spanish island of Menorca, of which Mahon is the capital. Called "Xoringuer" (pronounced: sho-ri-gair), it is produced using a direct flame alembic still, beginning with a wine-based spirit that has been flavored with local juniper, and then aged in American wooden barrels before being bottled.

Vilniaus Džinas or Vilnius Gin: produced in the capital city of Lithuania, Vilnius, this London Dry style gin, of fairly recent production (when compared to the other similar PGIs) starting in the 1980s, is made with juniper, dill and coriander seeds, and orange peel.

Mezcal Gin of Pierde Almas: produced in Oaxaca, Mexico, with a base distillate of double-distilled mezcal and featuring nine botanicals, which are macerated for 24 hours, then distilled a third time all together, resulting in a perfect London Dry Gin style. The botanicals used are juniper, coriander seeds, star anise and fennel, orange peel, cassia bark, angelica and orris root, and finally nutmeg. Bottled at 45% vol.

Steinhäger: this German product is from the Westphalia region, where there were once more than twenty distilleries in operation. Unfortunately, today most of them have closed, yet a few remain, the best known and widespread of which being the historic H. W. Schlichte, founded in 1766. This gin starts with what is called a juniper lutter, a sort of juniper fermentation that is subsequently distilled and blended with a neutral grain spirit, then finally redistilled. After this, it can be flavored again with juniper. Its traditional clay bottle is called a "Crucca".

Kraški brinjevec stà, Karst gin or Karst juniper brandy: also known as "Gin Kraški", this gin is produced in Slovenia between the Karst and Brkini regions, obtained from the fermentation (at least a month duration) of local juniper berries, distilled twice in a copper pot. Also derived from this distillate are both the Serbian Klekovaca, a type of plum brandy aromatized with juniper berries, and the more well-known Borovicka, also known as juniper brandy, produced exclusively in Slovakia. The origins of Borovicka date to the 16th century, and its aroma must derive principally from juniper, yet along with this, another type of traditional botanical can be used, from a tree native to the Mediterranean known as "Cade" or "Cedar". This is one of the spirits protected by Regulation no. 110/2008 of the European Parliament. It must contain the following ingredients: a base spirit of grain and juniper, some "sweet" additions, and purified drinking water from Slovakia's High Tatra Mountains. Borovicka

can also contain juniper berries and sprigs, added to the spirit post-distillation for a touch of color. With a mill, the juniper is ground into a very fine powder, which is then placed in a fermentation vat along with yeast, nutrients and warm water. The must is then fermented and then distilled. The resulting product is subsequently used as a component of Borovicka.

Waragi: produced in Uganda, this gin's name means "war gin", a name bestowed on it by British expatriates in reference to a local spirit known in the Ugandan language as enguli. Historically linked to the English soldiers who raided East Africa and enlisted brigades of Nubian soldiers to support their military endeavors, this gin was invented to keep spirits high among the recruited men. In 1965, the so-called "Enguli Act" decreed that distillation of this spirit required a license, that distillers be obliged to sell their product to the governmental organization Uganda Distilleries Ltd, producers of branded bottled products, and that it be marketed as "Uganda Waragi" (a millet distillate, today produced entirely by East African Breweries Limited). When sold abroad, this product is always redistilled (at least once, sometimes twice), which differs from the version that can be drunk courtesy of distillers based in Uganda villages. The base ingredient of Waragi ranges from cassava and bananas to millet and cane sugar.

Ginebra San Miguel Gin: oddly, this is the most sold gin in the world, produced by Mandaluyong in the Philippines and owned by the San Miguel Corporation. In the Philippines, a country with the highest per capita gin consumption in the world, the vast majority of gin consumed is Ginebra San Miguel. In this region of the world, as is commonly known, the base spirit derives from sugar cane, and it would appear that the only botanical used is juniper. Ginebra San Miguel Gin was produced for the first time in 1834 by the Destileria y Licoreria de Ayala y Compañia, the oldest distillery in the Philippines.

FLAVOURED SEAPORT

TYPE	ALCOHOL CONTENT	TECHNIQUE	GARNISH
Sour spicy	15.6% abv	Shake & DoubleStrain	Lemon peel and fresh red pepper

RECIPE (15 cl)
3 cl ELEPHANT LONDON DRY GIN
2.5 cl Ancho Reyes Liqueur
2 cl fresh lime juice
1 cl flavored honey mix
A few drops of vanilla liqueur
Dried chilli pepper
Thinned with Scortese Ginger Beer
A few drops of egg white

PREPARATION
Honey mix
Stir together 300 ml honey and 200 ml warm water until completely blended (the usual proportions are equal to 40% water and 60% honey). Add a few lemon, orange and grapefruit peels, being careful to remove all the white bits from the peel, and vacuum seal. Let rest for 14–18 hours.

METHOD
Thoroughly chill a champagne glass, or use one that has been stored prior in the fridge/freezer. Cool the shaker perfectly (then toss out the ice) and then crush a small dry chilli pepper inside it (be sure to check the pepper's spiciness by using Scoville scale based on its origin, and also by tasting it, to avoid ruining the drink's balance). Pour in all the ingredients except the ginger beer. Add a few drops of fresh or dehydrated egg white. Shake vigorously. Double strain to remove all remaining floating bits (both ice and pepper) and finish with the ginger beer.

Chapter VI

The Gin and Tonic
and other stories

"I need to be myself. I can't be no one else. I'm feeling supersonic. Give me gin and tonic"

Oasis

Nothing is stopping you from drinking it neat, in small sips. Indeed, this is by far the best way to appreciate the work of the master distiller—the "father" of each and every gin—to perceive its nuances, the role and interaction of the botanicals, and the distinguishing features across production styles. In every case, however, the "mother" role is always the juniper plant. So, if you have come this far with this book, you will have no doubt by now that speaking about gin in the singular is entirely reductive, not just on account of the various organoleptic differences between historical gin types (such as Jenever and Plymouth), but also given the veritable creative explosion in the last decade that has influenced gin, a spirit with ancient roots and, at the same time, an eye towards the future. Gin's extraordinary success, however—its global renaissance, the ensuing expansion of labels and countless re-interpretations—would never have managed to spread across the five continents were it not for the men and women behind the bar, the musicians of gin, as it were, skilled in reading gin's score and interpreting the craft in infinite ways. Cocktails featuring gin, often as the protagonist and sometimes in a splendid supporting role, are so many that one could write an entire book ad hoc. Here, instead we will celebrate the three pillars of mixology: the gin and tonic, the Negroni and the Tiki blend, all three great contributors to the legendary status of this distillate so dear to us. The order is not really important, yet how could we not start with the immortal classic, the drink enjoyed by every generation and every age, the gin and tonic? And here you have it…

Herbal G&T

TYPE	ALCOHOL CONTENT	TECHNIQUE	GARNISH
G&T twist	10.8% abv	Built	Sprig of fresh thyme placed on top or hung from the edge of the glass

RECIPE (20 cl)
5 cl RIVO GIN
15 cl Scortese Pure Tonic
1 bsp thyme syrup
A few drops of celery bitters

METHOD
Chill a large tumbler glass very well, or store it in the fridge/freezer prior. If you chill the glass with ice, make sure you empty all of it out before adding the other ingredients. First pour in the bitters and the syrup, followed by the gin. Mix well. Pour in the tonic water. Lastly, carefully add a large chunk of slow ice. Garnish and serve with the small bottle of tonic alongside.

PREPARATION
Thyme syrup

Place ½ cup of fresh thyme and 2 full cups of cold water in a small saucepan. Turn on the heat and bring to the boil, then immediately remove from the heat. Let cool, filter well with a fine mesh strainer, then add white sugar to obtain a syrup at 62 degrees brix.

NOTES
Serving this G&T with a straw is strongly discouraged.

Gin and Tonic, seems simple enough...

Now let's head back to school, to the little school desks we all remember, to discuss a bit of "literature". A very specific genre, in fact—those first sacred texts on mixology, books that laid the foundations for modern mixing, and those pages so many bartenders read long before the internet came along and that, even today, vital references for those in this field. These are: *The Bon Vivant's Companion* or *How to Mix Drinks* by Jerry Thomas (1862), *Bartender's Manual* by Harry Johnson (1882), *The Modern Bartender's Guide* by O. H. Byron (1884), *The Flowing Bowl* by William Schmidt (1891), *Modern American Drinks* by George J. Kappler (1895), *ABC of Mixing Cocktails* by Harry McElhone (1921) and *How To Mix Them* by Robert Vermeire (1922). Naturally, we will be citing only the most important of these. How many of these books include something on the gin and tonic? Not even one. The first book to ever mention this drink was *The Artistry of Mixing Drinks* by Frank Meier, published in 1936, which could lead some to infer that the gin and tonic must have first appeared on the scene in the late 1920s. This, however, doesn't mean the cocktail did not exist prior to that period. Rather, the G and T was simply regarded as more of a healing drink, until the period known as the Golden Age of cocktails, when this particular drink received its honorable status.

That said, we absolutely would never wish to belittle this drink's history. Instead, we should give credit for what it truly is, a drink born as a popular beverage and consumed in abundance. Much like today. So much so that if we tried to ask ourselves which is the most consumed aperitif in the world, the answer would be none other than the gin and tonic (in the world, that is; not in Italy). And not just because in Italy the tendency is to view the gin and tonic as a rival to other cocktails, but also because in Italy, gin is an ingredient in "our" aperitifs, and perceived as hard alcohol. This is a myth that we would like to dispel in the following few lines. Just as we would like to counter the notion that a gin and tonic is an easy drink to make. It certainly is, but only if you do not give it your best! If, on the other hand, your goal is a perfect gin and tonic, whether at home or while at the bar, allow me to share some rules I have developed over my many years of work in this area.

Firstly, given that this drink is prepared with tonic water, the first bit of advice I can give is to work exclusively with a tonic contained in glass or aluminum bottles. Plastic is an awful way to conserve CO_2 and in a short amount of time the internal pressure tends to disappear. Moreover, even when you do choose glass or aluminum, rotating your select tonics is still a good practice. No less important is keeping tonics cool by storing in the refrigerator, just as you should the spirit itself (or in the freezer). We realize that these practices present a series of difficulties for professionals. However, note that bottles stored on bar shelves, while perhaps esthetically pleasing as they rest there, illuminated, do not encourage conditions for achieving the best results.

Chapter VI

Third point: the glass. Unless you are dealing with tastings, for which small doses make sense, it's best to avoid using a low tumbler. Choose a large tumbler, one with a capacity of 28-36 cl (which by the way was the type chosen by 50% of respondents in a survey conducted on this topic). The same survey on the choice of a gin and tonic glass showed an excellent result (28% against and 50% in favor of the tumbler) for the balloon glass, copita (or sherry) glass, or red wine glass. Though technically uncomfortable and heavy, this glass has the advantage of a stem that allows you to avoid full-hand contact with the glass itself, something that should also be avoided in the case of the beloved large tumbler.

And the glass, just like the distillate, should be already cold when put to use. It can be kept in the freezer or, more easily, filled to the edge with ice for rapid cooling. When you begin to prepare the drink, the glass should be thoroughly emptied. At this point, you can pour in the dose of gin, followed by the tonic, being careful to proceed with delicacy, and tilting the glass so as to not let it "foam". Preparing a G and T by following this method, you avoid needing subsequent stirring to amalgamate the liquids. This is beneficial for two simple reasons: firstly, because the liquid of greater volume, poured after the one of lesser volume, facilitates amalgamation; and also because the tonic's density is always higher than the gin's, even a gin that has been kept in the freezer, and therefore the tonic passes through gin, blending the drink to perfection, naturally.

Up until this point, all of it seems rather simple to me. A couple things are still missing, however, if we wish to achieve what we like to call a "perfect serve". Before anything else, the most important ingredient in any cocktail is the ice. Our advice is to use a single chunk, decidedly preferable over multiple cubes, as the latter means more surface contact with the liquid, resulting in greater dilution than when using a single cube. You can also choose the double frozen version, but in this case you must be careful to let it breathe a little at room temperature because, if used immediately in a glass and in already cold liquids, there's a risk that the tonic water will foam up, as the non-tempered ice (with its dry surface) would act as a "magnet" for the CO_2 bubbles in the tonic. The ice should be of perfect quality, adequately dense and not punctured. Note, too, the importance of keeping the water's hardness constant by using a softener, of properly cleaning the inside of the ice machine and, lastly, of storing the latter correctly: in a cool dry place, within its storage container.

Here we will outline a brief yet useful digression on ice dilution, taken from an experiment by Giovanni Cecarelli and shared in his book *Miscelare* ("Mixing"), co-authored by Federico Mastellari. First 127.8 grams of light rum is poured into three identical glasses. In each of the glass, a differently shaped piece of ice is placed (cube, small cube, sphere), each weighing 105 grams, at a temperature of 0 °C. The only element that varies here is the surface area of the three shapes of ice: 251.6 cm^2 for the small cubes,

132 cm² for the cube, and 113 cm² for the sphere. Wait precisely two minutes and filter in due manner.

The cubes will have generated a dilution equal to 19.2 g, bringing the drink to a final temperature of 5.5 ° C. The small cubes will generate a dilution equal to 14.2 g, bringing the drink to a temperature of 7.9 ° C, while the sphere will have generated a dilution equal to 12 g, bringing the drink to a temperature of 9 ° C.

The conclusion? It is not the quantity of ice (as interpreted in weight and thus mass) that affects the speed of dilution and cooling but rather the surface area of ice that comes into contact with the liquid. The greater the ice-liquid contact surface, the faster the cooling and dilution happens.

The final detail to consider is the garnish. While to the consumer the garnish might seem no more than a simple decoration, for the barman it is an element essential to composing a perfect serve gin and tonic. The first recommendation we would like to make is this: resist using dozens of different garnishes, else you simply transform your drink into some kind of salad. The addition of citrus or fruit (or anything else) to the glass can be a "strong" signal; at the same time, remember that these flavors can significantly change a drink's flavor profile (in addition turning the glass into a small "aquarium", a very unfortunate effect indeed). So, our advice is to play it safe. Place the garnish on the edge of the glass, and allow the drinker to decide what to do with it. Moreover, the choice of a specific garnish must be made according to the gin being served. Each has a different aromatic profile, designed and obtained through the work of its master distiller, perhaps after several attempts. The task of a good barman is to enhance and elevate this profile, not change it or, even worse, to compromise or even destroy it. So the first thing to do is to taste the gin (or at least review its organoleptic profile carefully) to understand which aromatic notes predominantly characterize the gin you're working with. Only then can the most suitable garnish be selected. Specifically, fruity notes will call for raspberries, mangoes or blackberries, while floral notes lend themselves to roses, lavender or lotus leaves. Garnish ingredients should be washed prior, a practice that applies equally to lemon and grapefruit peel (but never orange peels).

As noted prior, the garnish should be positioned on the edge of the glass, where its aromatic notes will be pleasantly perceived by the drinker's retro-olfactory receptors when the glass is brought to the lips. Over the years, we have come to realize just how fundamental a role the nose plays in tasting a drink, and so it is also for this reason we feel obliged to make a case for always using fresh products that recall one or more of the drink's components, while leaving aside entirely the use of essences or perfumes. Finally, what would the gin and tonic be without tonic water? Until just a few years ago, a gin and tonic could essentially rely on only one type of tonic water. And in some cases (the worst of them, in truth), it was even poured from a spout. Nobody longs for that time, although the price to pay for its end is today's veritable explosion of choices when it comes to types of tonics. The tonic water explosion and its attendant organoleptic diversity, however, are factors that do help build up more solid foundations of the perfect serve concept. This is because tonic water in a gin and tonic, contrary to what one might think, is not simply an extra—rather, it takes the leading role in a cocktail. Let us consider it, for example, merely from the point of view of volume. In a perfect gin and tonic, the tonic water amounts to ¾ of the drink in terms of volume (that is, given a twenty centiliter drink, five cl of gin and fifteen cl of tonic). This also makes it easier to understand just how unfounded the (all-Italian) idea of a gin and tonic as a hard alcohol aperitif is. Using a gin with an average alcoholic content of 40% volume, the non-alcoholic contribution of the tonic would mean a drink at 10% volume: lower alcohol content than a glass of wine! And, if we factor in the inevitable dilution of ice in the glass (which obviously varies based on the quality of the ice, the consumption time, the room temperature, the gin and the tonic), it is easy to imagine a resulting drink of just over 6% volume—in short, little more than a medium-sized beer!

Having established this, the fundamental question remains: among the thousands of available tonics, which should be combined with your chosen gin? Here, the term pairing seems a bit forced, so perhaps it's better to speak of symbiosis, to stay within the concept of the perfect serve (my preference). Yet, to me it seems important to strongly emphasize a primary element: no static analysis scheme exists for the perfect combination in a gin and tonic. I agree having such a guide would be nice, yet several variables must be considered still, including human factors that, for the barman, naturally must refer to the tastes of the consumer. That said, we must always avoid letting the tonic override the gin, making certain it does not compromise the balance and equilibrium, essential requirements of perfect serve gin and tonic. I have therefore attempted to untangle this web of countless tonic waters on the market, trying to distinguish them organizationally, into categories I have outlined over the years. The first category is neutral tonics, tonics that could be defined as spontaneous, simple, natural. The traits of these are water, sugar, quinine, mineral salts and some citrus perception, which helps prevent upsetting balance when working with a delicate, light and not highly structured gin. Within this first category are the slim, dry or light tonics, an area that is growing significantly after the application of the so-called Sugar Tax in the United Kingdom on April 6, 2018.

Next come the aromatic tonics, each of which have their own identity, structure or personality, if you like. This type of tonic, when combined with a gin with a specific character, can work to elevate its identity. Finally, the fruity, citrusy and floral tonics, which are not represented by the group of tonics with fruity and/or floral notes, but rather by brands wishing to emphasize a very precise aroma—to such an extent that it can completely override the classic concept of tonic. An effective instructional example of this is Fever Tree Elderflower, a tonic in which the elderberry note dominates over the classic quinine one. Ultimately then, for me this is a category of tonic that is difficult to combine in a gin and tonic, on account of its dominating characteristic. My suggestion is to use this type of tonic in drinks such as Bucks, Collins, Fizzes or Cobblers.

Finally, another point to clarify regarding tonics concerns their effervescence, which we consequently also find in today's gin and tonic. With the exception of some natural sparkling waters—such as the famous German Selterwasser (a water rich in CO_2 that flows near Selters in the Taunus mountains, and from which the overused term seltzer derives) and the equally well-known Ferrarelle, Lete, Sangemini, Uliveto and the French Perrier—the artificial introduction of carbon dioxide in water is a practice owing to English chemist and philosopher Joseph Priestley. In 1767 Priestley discovered that by suspending a leather bottle filled with water over a fermenting vat of beer, bubbles would form in the water. He called this gas produced by yeast's action in beer "fixed air". A few years later, he published his book titled *Impregnating Water With Fixed Air*, in which he writes about replacing beer's fermenting action with oil of vitriol, or sulfuric acid that develops carbon dioxide when it comes into contact with gypsum. Thanks to Priestley's research and studies, he is considered the "father" of modern soft drinks.

Later it was discovered that effervescence could be achieved via the simple chemical reaction of tartaric acid and citric acid (both release hydrogen ions) and sodium carbonate (which takes in the hydrogen ions released from acids). The latter is then transformed into carbonic acid, which tends to then decompose, resulting in water and carbon dioxide: the precise gas responsible for effervescence.

Some fundamental steps led up to the creation of the tonic we are familiar with today. These include: the diffusion of artificially made sparkling water, the possibility of preserving its properties using containers such as the Torpedo Bottle (patented in 1809) or the dispensing soda fountain (patented in 1819 by Samuel Fahnestock, after its forerunner Charles Plinth's soda siphon in 1813), and the ways it could be flavored with various aromas, among them quinine. Soda fountains later gained extreme popularity thanks to John Mathews, considered the father of American soda water, who developed simpler and more efficient models in 1832. Their use spread to bars and street vendors. His invention consisted of a "metal lined chamber" wherein sulfuric acid and calcium carbonate were mixed to produce carbon dioxide. Subsequently purified, it was then passed through a cold water tank for about a half hour (to properly dissolve the gas) and finally was pumped towards the taps.

The ice chronicles

A brief yet worthwhile digression comes next, to discuss what should rightly be viewed as the most important ingredient of a perfect gin and tonic. No, despite the easier assumption that comes to mind, it is not, in fact, the gin. It's not even the tonic. That's right: it is actually the ice. Ice is the ingredient around which a skilled barman builds his entire workstation. Everything turns on the ice. And yet the historical journey leading to modern ice-producing machines has been a long and adventurous one, full of obstacles and failures. It has also been an exciting journey, with origins dating to ancient 1700 BC Persia, a time and place that saw the world's first iceboxes, the first attempts to preserve snow and ice as long as possible, after being gathered from nearby streams or lakes when conditions were good, or in the nearest mountains when conditions were less accommodating. These iceboxes, usually underground or in caves, were objectively expensive to build and maintain. In more fortunate cases, ice was supplied from nearby, while in other cases, from several miles away. Not surprisingly then, only nobles and the richest and most highly placed classes of people could feasibly possess an icebox.

It was not until the 18th century that the first refrigeration machines were born. In 1755, a Scottish doctor, chemist and academic named William Cullen designed the first refrigeration machine, using a pump to create a partial vacuum, placed over a container of diethyl ether, which he then brought to the boil to absorb heat from the surrounding air. The experiment worked, even creating a small amount of ice. But it had no practical application. Not long after, two Americans tried the idea out again. Benjamin Franklin and chemistry professor John Hadley collaborated on a project focusing on the study of the principle of evaporation, as a way to rapidly cool objects. The application was conducted on the bulb of a mercury thermometer, actually reaching a temperature of −14 ° C. Commenting on the experiment, Franklin wrote the following with a certain note of cynicism: "From this experiment, one may see the possibility of freezing a man to death on a warm summer's day."

In 1805, American inventor Oliver Evans described a vapor compression refrigeration cycle, using vacuum sealed ether, while in 1834 another American, Jacob Perkins, built the first functioning vapor compression refrigeration system in Great Britain. Scholars of the time, aware of its concrete applications, were highly intrigued by the notion of creating cold artificially, without resorting to natural ice. In 1842, an American doctor named John Gorrie was studying tropical diseases in Florida. Gorrie was, in addition to working on cleaning up area swamps and using mosquito nets at night to fight the disease, convinced some hospitals recognize the need to cool down the rooms housing feverish patients. Patented in 1851, Gorrie's ice machine worked, but it did not see widespread distribution, nor commercial success. Things went better for James Harrison, a British journalist who had emigrated to Australia. His refrigeration system, another example of vapor compression, used ether, alcohol, and ammonia. Patented in 1856, it found its practical application in breweries and meat distribution companies. In 1861, around a dozen of his systems were in use throughout Australia. In Europe, a turning point came along, in the form of a German engineer specializing in steam locomotives and a professor at the University of Munich. Carl von Linde had dedicated himself to the study of refrigeration between 1860 and 1870, inspired (also financially) by increased desire among important breweries of the time to produce an increasingly popular low fermentation beer, called lagers, all year round. Its 1876 patent made it possible to use gases like ammonia, sulfur dioxide and methyl chloride as refrigerants, widely used until the end of the 1920s. Ammonia was replaced by freon gas in 1931, having been synthesized for the first time by the Frigidaire company in 1930. It remained in use until almost the 1980s, when studies revealed its harmful effects on the environment—freon is considered one of the principal causes of holes in the ozone layer. Demand for this precious commodity continued to drive scientific discoveries and technological innovations, and the possibility of producing it artificially meant business activities offering enormous economic advantages. Proof of this can be seen in the story of Frederic Tudor, an American who lived between 1783 and 1864, known in the papers as "The Ice King". At the age of eighteen, Tudor realized that the ice trade would make his fortune. In 1806, he sailed his brigantine Favorite, in which he had invested his entire savings, from Boston to Martinique, with some 130 tons of ice on board. But upon arrival, Tudor realized there were no iceboxes on the island, and therefore the ice had to be sold very quickly. He refused the first offer he received, and as a result, lost the entire cargo. It simply melted. Tudor did not lose heart, however. The following year he managed to deliver and sell an ice cargo of one hundred and eighty tons in Cuba. In 1825 Tudor shipped over four thousand tons of ice by sea, thanks to an improved system of isolation in the ship's holds and the construction of evermore ice boxes at the landing points. Together with his partner, Nathaniel Jarvis Wyeth, Tudor also studied a system for cutting ice into transportable blocks and, in 1833, set about accomplishing his greatest feat: delivering two hundred tons of ice to Calcutta, India. The sixteen thousand-mile journey took six months. An incredulous crowd gathered to meet the ship at the port on the day of its disembarkation. Before their eyes was a ship from whose hold streams of water poured (the total loss of cargo reached eighty tons). Those fortunate enough to touch one of the ice blocks with their hands felt a burning sensation. But the deed was done, and from that moment onwards, the wealthy gentlemen of India could no longer do without Tudor's ice. At least until 1878, when it was replaced by the Bengal Ice Company.

If Tudor is emblematic of a bygone period, other stories of a time when ice was literally worth as much as gold (or nearly so) deserve to be told. A significant example is the story of Jacob Hittinger, an American who obtained ice from a lake in Massachusetts and shipped it to London, along with two American bartenders who explained how to prepare drinks like the cobbler, julep, smash and others. And then there's Henry Colman, a man who in 1845 opened a warehouse on the London Strand, from which uniformed porters left and delivered the ice on the doorstep.

Life-saving quinine

Malaria that affects humans is transmitted exclusively by mosquitoes of the genus Anopheles. Already widespread in ancient times (doctors such as Hippocrates and Galen cite it), malaria was a veritable scourge for centuries, if not millennia. According to a report by the World Health Organization, between 350 and 500 million people are still infected every year, with about one million deaths attributable to this cause.

For more than two centuries, malaria was treated with bark from the cinchona, a tree native to the pre-Andean area and that grows throughout the forests of Colombia, Peru, Bolivia, Ecuador and Venezuela. It contains alkaloids, the most important of which is quinine. In areas affected by malaria, daily doses of quinine to treat the infected was common. It had, however, one disadvantage: the highly unpleasant flavor. To offset the taste, in the early days of quinine, it would be mixed with water, sugar and lime, essentially becoming what could be considered the forerunner of all tonic waters.

A historical anecdote tells us that in 1638, Francesca Henriquez de Ribera, a noblewoman and wife of the Peruvian viceroy Luis Jeronimo de Cabrera, was dying. After several fruitless attempts to save her life by relying on the official medicine of the time, a Jesuit priest managed to heal her with a dose of a local medicine called "Ayac Cara", also known as "Quinquina" or "bitter bark". We can never know if this story is true or legend, but it is certain that the disease in question was malaria and the cure was quinine. Another compelling story is that of Agostino Salumbro, a Spanish Jesuit stationed in Lima. In 1631, he sent some cinchona bark to his headquarters in Rome, and just ten years later it seems certain that quinine was being used to treat endemic malaria in the swamps found south of the city.

The discovery of the properties of quinine, which not only calms fever and muscle spasms but also blocks the parasite's metabolism, leading to its death, was decidedly revolutionary. Imported to Europe (for the first time in Spain, soon to become its commercial hub), it was introduced as a mandatory medicine in areas at risk of malarial contamination and on ships. In 1803, Admiral Horatio Nelson, future victor at Trafalgar, ordered it be consumed in a mix with a dose of wine or rum by all sailors who wished to go ashore. And he had every reason to do so, considering what happened just a few years later to an English expedition landed in Walcheren, Holland, in an attempt to support the Austrians against Napoleon. An operation that ended in disaster, with over twelve thousand men infected by malaria. Since that time, British commanders would not go without quinine.

This substance has cured all manner of men, from sovereigns such as Charles II of England and Louis XIV of France, to soldiers, bourgeois and commoners. In 1901, the Italian state distributed some thirty tons of quinine among its population of thirty-two million inhabitants.

Cinchona bark soon began to generate commercial income, in a manner similar to substances usually considered more precious, leading to cultivation of the tree wherever climatic conditions were favorable. For example, in 1883 the island of Ceylon boasted a plantation growing around one hundred twenty-eight million cinchona trees, sufficient to supply all of Great Britain alone. Isolated in pure form as early as in 1820 by Pierre-Joseph Pelletier and Joseph Bienaimé

Caventou, two French chemists who also managed to isolate chlorophyll, strychnine and caffeine, it was then synthesized in the laboratory for the first time in 1943.

In 1858, the Englishman Erasmus Bond discovered a way to use quinine in a somewhat more pleasing manner, creating what can be considered the prototype of tonic water, containing quinine, salicine and bitter orange. He registered his Improved Aerated Liquid and presented it for the first time in 1862 at the London International Exhibition. His compatriot Thomas Whiffen was the first to establish himself as a large-scale quinine producer, and in 1870 he created the Indian Quinine Tonic, designed for British people living in overseas colonies, for whom daily doses of quinine were routine. To enjoy the substance better, they consumed it together with lemon or lime juice and a teaspoon of sugar. What's more, in British India at the time of Queen Victoria, the habit among colonists was to mix it with gin. Schweppes Quinine Tonic was also a great success, especially so when it could be mixed with a new type of gin hailing from the motherland, London Dry. Drunk as a daily ritual, on sultry afternoons in Bombay or Delhi, the drink even gained a name in the local Hindi language: "Chota-Peg". Reasons for its success were likely several, in part due to the pleasant taste of the drink, but also for reasons related to the properties of its various ingredients: the vasodilating, relaxing and euphoric effect of the gin; the anti-scurvy, disinfectant actions of lemon and lime; the antimalarial, calming properties of quinine; and the energy-producing effects of sugar. Note that here we are talking about a rather primitive gin and tonic, "flat" and invariably drunk at room temperature, containing quinine in "medicinal" doses when compared to today's regulated quantity of the substance when used in its natural form: maximum 83 parts per million (as a treatment for malaria, the current daily dose is instead set between 500 and 1000 parts per million).

This was a gin and tonic light years away from the elegant, harmonious cocktail it would become, still yet the die was cast.

Bitter is better

Around the same time as tonic waters were being developed, another future protagonist of mixology was taking its first steps. Like gin, bitters were born primarily for healing purposes. While daily doses of quinine were beneficial in the treatment of malaria, another common disease at the time, gout (described in the 17th century by Englishman Thomas Sydenham as a consequence of "ease, voluptuousness, high living, and too free a use of wine and other spirituous liquors") came to be treated with a digestive "concoction" called bitters. A sort of proto-gin made with watercress, horseradish, wormwood and angelica root, Sydenham's very popular Bitters soon became a successful remedy for this disease. In 1712, thanks to Reverend Richard Stoughton, one of the most famous products in this area was born: Stoughton's Elixir, containing twenty-two ingredients and recommended in doses of 50-60 drops in a glass of

THE SCHWEPPES "MOVE"

Johann Jacob Schweppe was born in Witzenhausen, in the German region of Hesse, on March 16, 1740. A silversmith and jeweler from Geneva, as well as a shrewd inventor and entrepreneur, Schweppe quickly became passionate about Joseph Priestley's research on carbonization. His interest led to the invention of a true and proper system for adding carbon dioxide to water (1783). His Geneva System or Geneva Apparatus consisted of a machine that enclosed an agitator capable of generating carbon dioxide from a mixture of gypsum and sulfuric acid. The gas passed through water and, with the help of a pump, was passed to a wooden container and mixed (as suggested by Priestley). Following the results of Priestley's research and his soda water, this invention gave birth to the first Schweppes Water, flavored with roses, flowers, roots and more...

In 1790, Schweppe partnered with Paul Nicholas and his father Jacques Paul, also involving a well-known pharmacist from Geneva named Henry Albert Gosse. The partnership of Schweppe, Paul and Gosse enabled them to work for the first time with distilled water, establishing other important collaborations. On January 9, 1792 Schweppe moved to London, where the presence of carbonated waters was already well established, though inferior to those crafted by Schweppe. Still, his success was slow in the making. After a year, his partners asked Schweppe to close down operations, but he would not not give up. On February 20, 1795, the partnership dissolved, and meanwhile, a year earlier, Schweppe was to meet Erasmus Darwin, grandfather of the

water (sweetened or not), beer, white wine or a dram of brandy. The first drug to obtain the British Royal Patent, the elixir remained a top selling product for a long time. Alongside the talented Sydenham and Stoughton, however, countless charlatans were also to launch their products within the bitters arena, individuals who have gone down in history under the derogatory label of quacks. These remedies, inexpensive but useless or possibly even harmful, often contained strychnine and went by several different names, such as Daffy's Elixir, Godfrey's Cordial, Bateman's Pectoral Drops, and so on... The work of the Italian Giovanni Maria Farina was different. A perfumer who had successfully exploited the basic principles of medicine and the characteristics of these proto-gins, Farina created a new fragrance more famous Charles, who became his lawyer. Thanks to knowledge, corporate connections and high-ranking support (and also due to the fact that soda was prescribed as a treatment for kidney stones), Schweppes Water and "soda water" soon became household terms. Additionally, this time also saw the end of bottling soda in wooden containers and the start of circulating the product in glass containers, which better maintain pressure. Schweppe continued to work diligently on marketing his product with highly effective advertisements. His soda water was soon being drunk everywhere. Doctors began prescribing it to those suffering from debilitating disorders, nervous afflictions, fevers and kidney problems. As a testament to its success, the British government decided to tax every bottle produced with an excise duty of three coins. While the tax remained in force only until 1840, the Schweppes brand is still very much with us today.

in 1709 using bergamot, lemons, neroli, oranges, rosemary and cardamom, infused in a neutral distillate. He baptized his new fragrance Eau de Cologne in honor of the German city where he worked. In 1783, a French officer named Nicholas Husson began selling a bitter whose ingredients included saffron extract. Only later was it discovered that saffron contains *Colchicum autumnale*, a useful remedy for gout.

In 1868 John Rack published his book *French Wine and Liquor Manufacturer*, outlining eight recipes for bitters to be used for medicinal purposes; however, the two main products destined to go down in history, products stilled considered today the cornerstones of the genre, go by the names Peychaud and Angostura.

For the first, we owe thanks to Antoine Amédee Peychaud. Originally from Bordeaux, and born into a wealthy landowning family in Haiti, he fled the island after the 1804 revolution, arriving in New Orleans aboard the schooner Brisk at the age of twenty-one. Here he later became a talented pharmacist, with a drugstore located in the heart of the French Quarter, at 437 Royal Street, where he would welcome customers with taste of his bitters—water, sugar and a dose of French brandy. A sort of rudimentary old fashioned toddy. Still on the market today, Peychaud's Bitter was so successful that by the time Peychaud died (in 1883), his bitters had completely supplanted Stoughton's Elixir in sales.

Even more famous than Peychaud, however, was, and is, Angostura Bitters, whose story begins in the early 1920s with the Prussian doctor Johann Gottlieb Benjamin Siegert who invented it. Siegert was in the service of Simon Bolivar, the man also known as the legendary El Libertador for his successful military campaigns in favor of the independence for Colombia, Venezuela, Bolivia and Peru. Having studied a drug that could treat dizziness, fevers and diseases like cholera, he then created a recipe using different types of herbs and spices, including *Cusparia Febrifuga* from Venezuela along with Galipea Officinalis. The first of these was also known as Angostura Barak, from the city of Angostura (bottleneck) located on the Orinoco River, today Ciudad Bolívar. Though it went by other names along its journey, including Amargo Aromatic and Siegert's Tonic, it is the name Angostura that we have inherited. As years passed, Siegert devoted himself exclusively to the production of his bitters, with his sons Carlos and Alfredo by his side. It was they who expanded the family business, by first moving it to Trinidad to escape uncertainties of the South American political atmosphere of the time, and then by bringing the product to Europe, where it became a huge success. At the beginning of the 20th century, Angostura Bitters could claim the status of official supplier to the royal houses of Spain, France and England. The company changed hands for a short time, then later returned to the Siegert family. Today the brand is present in over one hundred forty countries, and the name Angostura is among the best known globally, both in the bitters sector and also that of rum, thanks to their successful diversification in the aftermath of the Second World War.

While not nearly as popular or widely known as Angostura (which by the way also deserves credit for bringing one of history's first cocktails to life, namely Pink Gin), another brand of bitters worth mentioning as we conclude this section is Boker's. Produced first in New York, Boker's gained its fame as the favorite bitters of Jerry Thomas, the "professor" unanimously considered the "patriarch" of subsequent generations of bartenders, and who repeatedly mentioned this product in his publications.

YEARS TO BE REMEMBERED

1824 – Johann Gottlieb Benjamin Siegert's recipe for Aromatic Amargo appears on the market for the first time. Later it will take the name Angostura Bitter, by which it is still known today.

1856 – The term "mixologist" appears for the first time in the New York magazine The Knickerboker.

1858 – In London, Erasmus Bond patents his Improved Aerated Liquid, the first tonic water to be officially registered with a patent.

1871 – In New York, William Hernett registers his patent for history's first shaker.

1874 – English businessman William John Barritt opens a convenience store, selling self-bottled soda drinks in the back of the shop.

1888 – On January 3, Marvin Stone patents his spiral winding process for the production of the first paper drinking straws. Formerly a manufacturer of cigarette papers, Stone had one day wrapped a piece of paper around a pencil, sealing it with paraffin. In 1906, the Stone Straw Corporation created the first machine for the industrial production of spiral straws, while the "accordion" spiral straw came later, invented in 1937 by Joseph Friedman, in San Francisco.

1892 – William Painter invents the crown cork.

1904 – Canadian pharmacist John McLaughlin creates his pale ginger ale, patented three years later and destined to become the celebrated Canada Dry Ginger Ale.

PERFECT BALANCE G&T

AROMATICS
Coriander
Marshmallow
Elder flower
Calendula
Hibiscus
Rose
Vanilla
Honeysuckle

FRESHNESS
Lavender
Chamomile
Violette
Makrut lime leaves
Lemon Grass
Mead wort
Lemon balm

HERBACEOUS
Earl Grey
Sencha tea

RESINOUS
Spruce
Pine
Juniper

VEGETAL
Thyme
Rosemary
Cucumber
Geranium
Cannabis sativa
Mugwort

REFRESHING
Mint
Sage
Eucalyptus
Basil

FLAVORSOME
Woody
Briny
Smoked
Fennel
Ginger

WARM SPICES
Cinnamon
Cloves
Saffron
Coffee
Chocolate

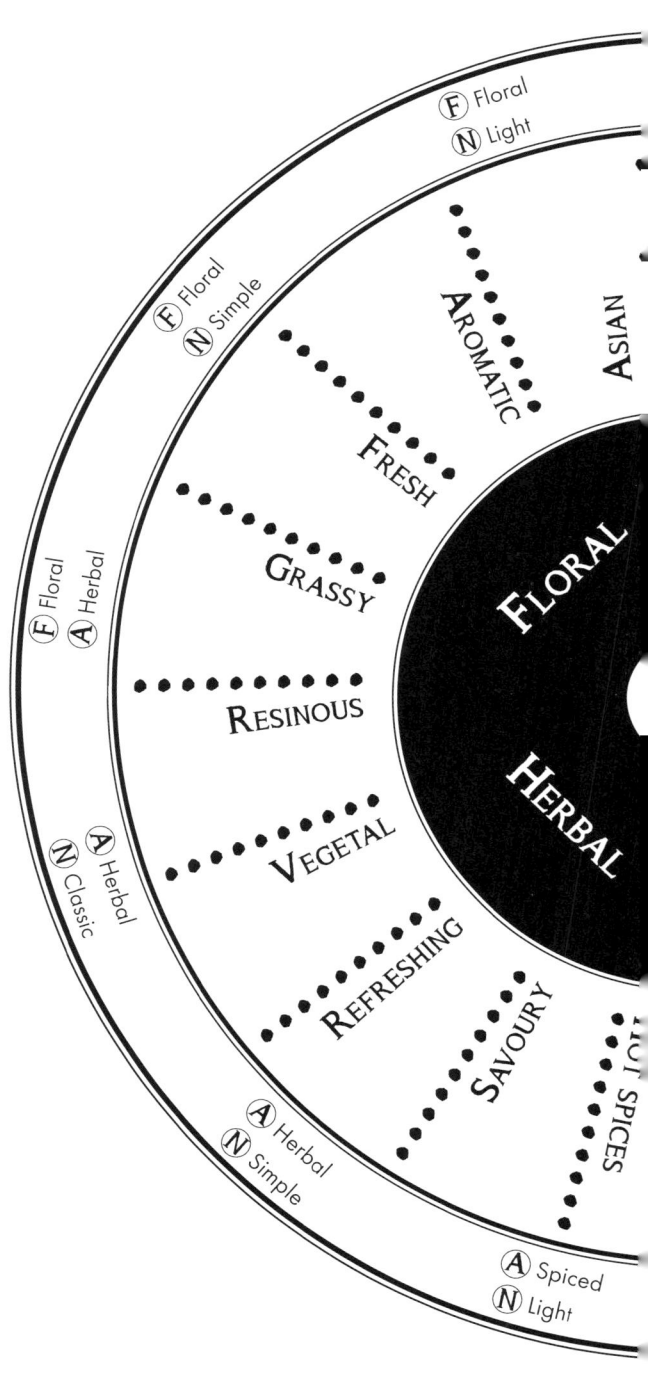

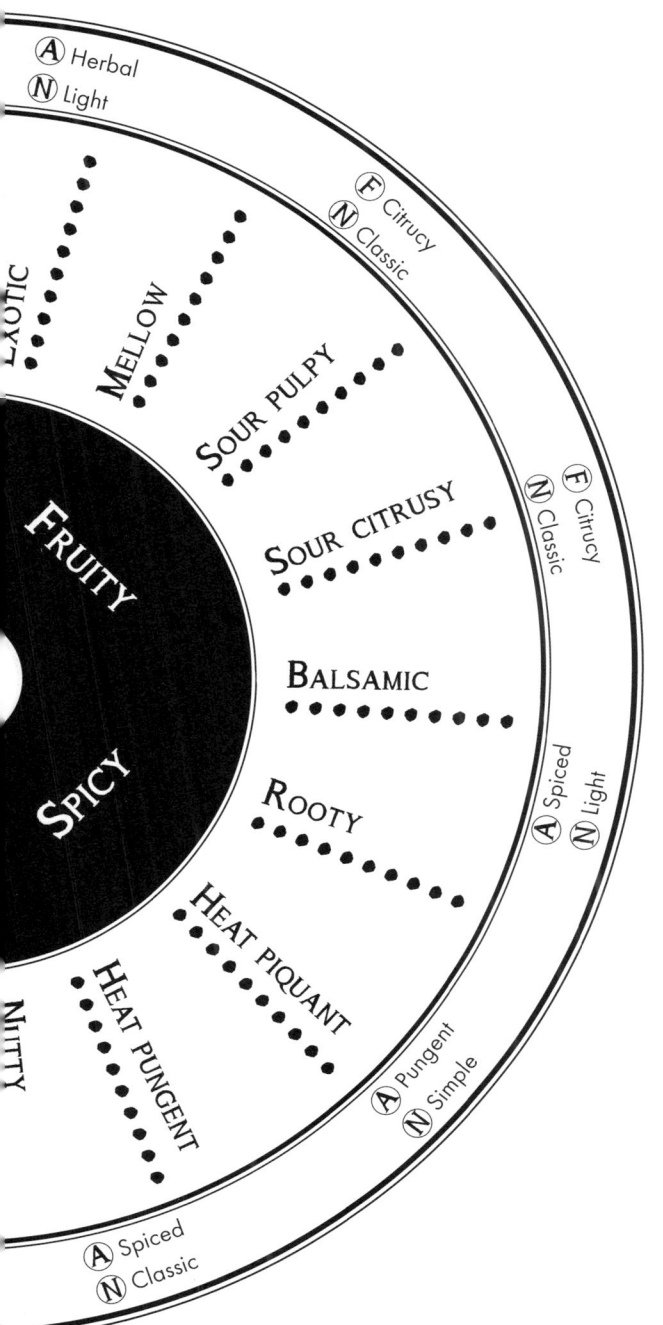

Fruit nuts
Nutmeg
Almond
Hazelnut

Warm sensations Pungent
Ginger

Warm sensations Spicy
Chili pepper
Pepper

Citric acids Pulp
Cherry
Raspberry
Red currant
Grapes

Citric acids Citrus
Lemon/Lime
Yuzu
Bergamot

Grapefruit
Vinegar
Buddha's Hand

Fruits
Apples
Pears
Peaches
Bananas
Melons

Exotic fruits
Pineapple
Passion Fruit
Mango
Feijoa
Coconut

Asian fruits
Lychee
Lotus flower
Guava
Papaya
Dragon Fruit
Durian

NEUTRALS

Light/Dry/Slim

- Cortese Light
- A. Q. Dry
- Fentimans Light
- Franklin Light
- T. Henry Slim
- The Duchess
- Syndrom Sugar Free
- Fever T. Light
- Goldberg Dry
- Double Dutch Skinny
- Fentimans Dry
- J. Gasco Dry
- Royal Bliss Zero
- Doctor Polidori
- PHI Tonic Zero
- GlacierFire #1
- Canada Dry Diet Tonic
- Baker and Quinn Light

Classic

- Cortese Pure Tonic
- Brillante Recoaro
- Schweppes Classica
- Fever Indian Tonic
- A. Q. Monaco
- Schweppes Premium
- Goldberg Original
- Russell&Co
- London Essence
- J. Gasco Indian
- 6 O'Clock
- Erasmus Bond
- Franklin Indian T.
- Britvic
- Seagram's
- Peter Spanton N.1
- Royal Bliss
- Monelli
- Double Dutch Indian
- Abbondio Dirty Soul
- BTW
- Syndrom Raw
- Bickford
- ABBAS Blu T.W.
- Levico Bio T.
- 3 cents T.W.
- Canada Drt T.W.
- Fusion Blue Rivers
- Devon Luscombe T.W.
- Quina-Fina T.W.
- SVAMI Indian T.W.
- Belvoir Indian T.W.
- EAGER T.W.
- Lamb&Watt T.W.
- Barker and Quinn T.W
- FairyQueen T.W.
- East imperial Old World Tonic

Simple

- Cortese 1959 Bio Originale
- Acqua Tonica Paoletti
- S. Pellegrino
- Vichy Catalan
- Plose Acqua Tonica

AROMATICS

Spicy/Pungent

- Cortese Strong
- Schweppes Pink Pepper
- GlacierFire #2

Herbal

- Scortese Pura Tonica
- Fentimans
- Fever Tree
- Mediterranean
- Q Tonic
- Fentimans Botanica
- Fentimans Connoisseurs
- Le Tribute
- London Essence Grapefruit & Rosemary
- Ledger's Tonic
- Peter Spanton N.9 Cardamom
- A. Q. Herbal
- Doctor Polidori Cucumber
- Gent Swiss Roots
- Indi
- Abbondio
- Mediterranean
- Markham
- The Duchess Greenery
- Syndrom Velvet
- East Imperial Royal Botanic Tonic
- GlacierFire #5
- 3cents AEGEAN
- Lamb&Watt Basil T.W.

Spiced

- Tassoni Superfine
- East Imperial Burma
- Erasmus Bond Botanical
- Ledger's Cinnamon
- Ledger's Licorice
- Franklin Rhubarb
- The Duchess Botanical
- Peter Spanton N.4 (chocolate)
- TauTonic 95
- GlacierFire #4
- Smoked EAGER T.W.

FLAVOURED

CRUSTY	1724 Original Yuzu Goldberg Yuzu Original Citrus Fever Clementine	Ledger's Mandarin Fraklin Pink Grapefruit Peter Spanton N.5 Lemongrass Royal Bliss Yuzu	Syndrom Grapefruit Tonic Acqua Tonica di Chinotto Lurisia East Imperial Grapefruit Tonic	East Imperial Yuzu Tonic Devon Luscombe Cucumber T.W. SVAMI Grapefruit T.W.
FRUITY	Original Berries Original Cherry Thomas Henry Cherry	Blossom Doctor Polidori Grape Double Dutch	Pomegranade & Basil Doble Dutch Canberry Double Dutch	Cucumber & Watermelon GlacierFire #6
FLORAL	Schweppes Orange Flower & Lavander London Essence Bitter Orange & Edelflower Schweppes Ginger & Cardamom	Fever Edlelflower T. Henry Edelflower Schweppes Matcha Tonic Abbondio Sambuco	The Dutchess Floral Syndrom Sweet Roses ABBAS verde Devon Luscombe Cucumber T.W.	Lamb&Watt Basil Ibiscus T.W. Barker and Quinn Hibiscus T.W. SVAMI Rosemary T.W.

FINDING THE PERFECT FLAVOR PAIRING FOR YOUR GIN&TONIC

1. Taste your gin neat, letting it linger on the first inner circle.

2. Identify the aroma that is most perceptible to you (for example, FRUITY).

3. Try to analyze it and break it down (FRUITY, in this case), relying on your own flavor experiences to recognize its qualities, and following the indications on flavor, such as Asian, exotic, mellow etc … Identify the most characteristic and score it from 1 to 10. Repeat this process with the others, if present. Example: FRUITY – exotic 8, mellow 4, sour 1, and so on.

4. Repeat this with HERBAL, SPICY and FLORAL.

5. When you are done, join the different points with a straight line, to make a sort of polygon formed by the various points.

6. Identify the most pronounced points, which will be indicated by the labels in the last circle.

7. Those labels are references to the type of tonic water that can be served with your gin, selected from those listed in the table on this page.

One circle, for example, indicates the category of AROMATICS, while the neighboring word (e.g. "Herbal") represents the group from which to select your tonic water.

The Negroni, Gin the Italian way

Interview with Luca Picchi, Negroni World Ambassador

Luca Picchi is the great keeper of all Negroni secrets. A Tuscan like Count Camillo Negroni himself, and a highly experienced barman, today Picchi is at the helm of the Florentine Caffè Gilli 1733. Yet he is also a tenacious researcher and a passionate educator. His book, *Negroni Cocktail. An Italian Legend*, condenses all there is to know about this famous drink into two hundred pages, while also exploring the fascinating historical setting and two lead figures involved in the Negroni's origins tale: Count Camillo Negroni, and the barman who materially invented this by-now legendary drink, Fosco Scarselli.

Luca, you are considered the greatest expert on the Negroni cocktail, whose 100th anniversary was in 2019. Can you tell us the secret of its remarkable success, which has lasted for over a century?
I believe that its long-lasting success is due to its overall simplicity. This is a simple cocktail to prepare, and moreover it bears a memorable name and is easy to pronounce in all languages. Much like the Martini, after all... And history also contributes to its charm: the popular Negroni emerged from a world war unscathed, and rose again in the 1950s to become the "welcome drink" in the Rome of la Dolce Vita, beloved by American artists who visited the Eternal City, by writers, and by the so-called "vitelloni" (Fellini's film and a term used generally to describe immature, driftless young Italian men). It is from that moment, I believe, that Negroni's global success took hold, which we were witness to on the occasion of its 100th birthday—a splendid achievement for this most aristocratic of cocktails.

The idea of adding a dose of gin created the magic. Do you think the same result would have been obtained with another distillate?
Surely not. Brandy, rum, vodka or other spirits, even if they had been present on the prestigious shelves of Florence's splendid cafes in the 1920s, would not have created this same "magic". The choice of gin was desired perhaps because Camillo came from an Anglo-Saxon background, having an English mother (Ada Savage Landor), but also certainly speaks to his experience of living in America in the early 20th century, a period that history would come to call the "Golden Age of Cocktails"—all likely contributing factors. Remember, too, that the Negroni was somewhat grafted onto the famous Americano cocktail, which exploded in Florence in the

wake of the arrival of numerous Piedmontese who left Italy's then-capital city of Turin for Florence in 1865. It was they who introduced the Florentines to their "Vermutte" and the "Vermouth hour". Thus, it was a short journey from Vermutte to Americano...

Is the idea of adding gin attributable to Count Camillo's request or to barman Fosco Scarselli's decision?
Adding gin was certainly a request from the count, and not an initiative taken by the young bartender Scarselli who, as journalist Marco Mascardi has noted, was simply its "first unsuspecting mixer".

The Negroni is drunk everywhere now, and it's a classic cocktail that can also be prepared at home. However, making a perfect Negroni is not that easy. Can you provide some suggestions on how to avoid making even the slightest mistake?
Firstly, your ingredients must be good quality, considering that Campari tips the balance here. Obviously, you can make a Negroni using any brand of ingredient, though I recommend the traditional labels. Otherwise, you will need enhanced taste skills and experience to achieve the "perfect balance" of aromas and flavors. The ice must be crystal clear. Having a good technique certainly helps, but you can go far with a nice old-fashioned vintage glass to create an evocative effect that will foster full enjoyment of your Negroni.

Lastly, alongside the classic Negroni is a considerable number of variations. Do you have a favorite Negroni? And if your favorite is the classic preparation, do you have a "second favorite"?
The Negroni, like the Martini or the Manhattan, is a classic, simple drink. At the same time, these cocktails are also easily "twistable", meaning they can be modified or customized, if you will—a not insignificant detail to which much their enduring fortunes can be attributed. The important thing, in my opinion, is that the new variation of the drink does not distort its soul of the original, nor its structure or essence. My new creation is called Negroni Azteco, and it's a bit of a gamble, much like the famous Negroni Insolito ("Unusual" Negroni) was. Firstly, use a previously frozen cocktail glass and sprinkle the "cone" of it with a quality cocoa powder.

The doses: 20 ml of gin, 20 ml of Etna Bitters, 20 ml of Barolo Chinato Cocchi, 15 ml of light cocoa cream.

Stir according to technique and decorate with a twist of lemon peel.

"Unusual" Negroni

TYPE	ALCOHOL CONTENT	TECHNIQUE	GARNISH
NEGRONI twist	27.8% abv	Stir&Strain	Orange peel

RECIPE (6 cl)
1.5 cl STAR OF BOMBAY GIN
1.5 cl Cask Tales Campari bitters
1.5 cl Cocchi Americano
1 cl China Clementi liqueur
Drops of orange bitters
Coffee beans

METHOD
The method recalls a classic Negroni not made directly in an old fashioned glass but rather mixed prior in a mixing glass. After having sufficiently cooled your tools and drained the excess water, pour all the ingredients into the glass except for the coffee beans. Be sure to stir slightly longer in order to leave room for a slight dilution. Lastly comes the technique that gives all the characterizing notes to this "Unusual" Negroni: pour it into an iced cocktail glass through a small colander with coffee beans inside.

NOTES
This is a personal recipe created by Luca Picchi, Negroni World Ambassador and author of the book *Negroni Cocktail. An Italian Legend.*

Belle Époque

TYPE	ALCOHOL CONTENT	TECHNIQUE	GARNISH
NEGRONI twist	25.6% abv	Stir&Strain	Grapefruit peel

RECIPE (9 cl)
2 cl OXLEY London Dry Gin
1 cl China Clementi Liqueur
3 cl Carpano Antica Formula Vermouth
2 cl Bitter Campari, flavored with chamomile, lemon balm and grapefruit peel
1 cl soda

PREPARATION
Flavored Bitter Campari
Immerse the flavoring substances in the Campari for a period ranging from 6 to 18 hours, depending on the desired intensity. Filter first with a fine mesh strainer and then, when the cocktail is very cold, with an aeropress directly into the glass.

NOTES
A recipe by Samuele Ambrosi, included in the Negroni Cocktail Hall Of Fame in Luca Picchi's book, *Negroni Cocktail. An Italian Legend*.

METHOD
Unlike a classic Negroni prepared directly in the glass, here I suggest chilling an old fashioned glass very well with ice, but work directly in the well-iced mixing glass, pouring all the ingredients into it and proceeding with a slightly longer stirring, to leave room for a slight dilution.
For a perfect serve, which I sometimes use for this drink, I will stir it with the coarsely filtered flavored Bitter Campari, so that it retains some of the residue from the maceration process. Next, I pour this into a perfectly labeled bottle, stored at 4 ° C. To serve, I use a low tumbler with a single chunk of ice, the very cold bottle and an aeropress. The customer then places the instrument on the glass, pours the contents of the bottle into it, and proceeds to the final filtering. In this way, the drinker is involved in the execution of the final drink.

Gin in America and its role in Tiki bars

By Gianni Zottola, international expert on Tiki mixing

The history of gin in the United States has undoubtedly deep roots. Gin has always featured in the oldest mixed drinks here, circulating among those who invented and exported the "cocktail" throughout America and thus influencing the concept of mixing. Historically, gin consumption dates to the first English colonies, as one might easily guess, and continued to thrive long after and in spite of the subsequent American independence from England.

Very likely, gin's influence on American drinking culture is largely due to Prohibition in the 1920s, when the illegal production of this spirit flourished.

Producing a good whiskey, rum or cognac was complicated, given both the considerable technical difficulties and scarcity of raw materials. It's no surprise then that gin quickly became the easiest product to use as a substitute.

While famous gin cocktails like the South Side, enjoyed by Al Capone in his club, The Four Deuces, were certainly prepared with genuine, likely excellent gins that had been illegally imported, the same cannot be said of those gin cocktails consumed in basements or back rooms of shops transformed into popular speakeasies. In these places, approximate infusions of botanicals in large bathtubs could turn even an abominable into something sophisticated.

They were usually made at home by artisanal distillers, distillers that, in the case of Chicago, were often supplied by the organized crime syndicates that dominated the South Side.

Widespread throughout American cities, this "alleged" gin, better known as bathtub gin, was in essence a very bad alcohol masked by strong aromas, which in turn derived from spices stolen off from soda shop counters and from pharmacies. Though these gins certainly have in common with gins of higher production quality, or even the history of gin in the bigger picture, they do, however, play a fundamental role in the growth of gin's popularity in America.

With this in mind, it's not hard to imagine how the end of the Prohibition era coincided with significantly increasing imports and the overall diffusion of gin—this time of real, proper gin—as by that point it had become not only indispensable to the American palate; gin was now a truly important ingredient in most of America's traditional cocktails.

The end of Prohibition and the lead-up to reopening bars brought with it deep curiosity and interest in discovering other spirits. It is no coincidence that the genius work of Donn Beach (credited with opening the first Tiki bar, Don the Beachcomber) made his fortune thanks to rum, taking what was essentially a multi-faceted yet little-known distillate, one of little interest prior, and making it the exclusive distillate of exotic blending in the 1930s. In the years following, Beach's dominant rum-based mixing gave way to gin, which took up its new role in the latest cocktail type to arrive on the scene, Tiki drinks.

And Dry Gin and London Dry were also to see their own exponential rise in popularity and use around the mid-20th century. The Tiki and exotic blending expansion started in the mid-1950s and lasted for a decade, a period in which even the great masters of this drink style would not hesitate to serve cocktails using gin, or drinks with a blend of spirits in which gin's presence was notable. And since Tiki was the most influential and widespread mix ever, one can appreciate to what extent the market for gin expanded, reaching new levels of demand during this period, only in a more exotic form.

It was Trader Vic in particular, the man credited with promoting Donn's mix during the second Tiki era, who revolutionized exotic recipes, in line with these new, more modern standards.

Thus, a new way of drinking was born, one that adapted to the new American way of life, being undoubtedly lighter and more carefree. Within this context, gin's quality of freshness forms part of a simpler, more immediate mixing technique. The easy drinkability of a Trader Vic cocktail, for instance, is in most cases a far cry from the structural complexity of the rum-only cocktails that one could enjoy at a Don the Beachcomber's table.

With this new style, the United States definitively abandoned the concept of the static, strong, meditative cocktail, opting instead for the freshness of a quicker, easier and entirely more dynamic drink. This more evolved expression of the exotic cocktail was to benefit partly from the new use of spirits, gin in particular. And yet, although this type of cocktail is apparently simple on the palate, it is a simplicity achieved through the lengthy, ongoing and complex studies that definitively placed Trader Vic among the greatest bartenders of all time. It matters that he did so by taking inspiration from or trying to emulate Donn's mixing: the new style of mixing invented by Trader Vic was entirely innovative.

Gin's big leap into the world of Tiki cocktail mixing came along just when Trader Vic was opening one of his best venues in London, thanks to commercial agreements made with the famous Hilton hotel chain. Needing to meet the taste preferences of the English palate, Vic proposed mixing with local spirits, replacing Scottish whiskey with bourbon or rye, but also introducing London Dry and Plymouth as the base spirit of several drinks. He sometimes even substituted gin for rum drinks that called for an ingredient in the Spanish *ligero* ("light") or Puerto Rican style—essentially a cocktail already tending to be "light" on its own.

It should be noted that the gins used in Tiki mixing were predominantly London Dry, as they were certainly the most commonly appreciated type, but more specifically because their dryness worked rather advantageously when combined with a mix that tended towards acidic. At times this would be an extreme acidity, an acidity needed, though, to highlight aromas deriving from sweeter notes like pineapple, orange or syrup-based (for example passion fruit, grenadine, or the frequently used barley water).

In most cases, gin is used in the Tiki style, particularly when it came to the Trader Vic's mixing style, not so much with the goal aim of highlight or enhance itself (as happens for example in the powerful rums used by Donn Beach, which when incorporated into the entire mix are decidedly full-bodied, structured, complex), but rather to create instantly perceptible aromatic profiles, surprising and immediate, by using a distillate better able to enhance other components featured in the cocktail.

Unlike rum, gin, with its complex variety of botanicals, possesses a consistent aromatic profile, and it is definitely not a distillate that requires long or differing durations to open up and release its hidden esters and aromas, as can happen with an aged rum. Making it, in others, fully suited to the exotic drinking dynamic of the 1950s. And today, given the endless variety of gins on the market, there is no reason to not reinterpret these old gin-based Tiki cocktails with vermore original aromatic and flavor profiles (as long as the correlation with the older recipe's ingredients are preserved, of course).

Take this textbook example: for a cocktail that calls for cinnamon syrup, avoid using a gin that already heavily exudes this particular spice among its group of botanicals. Instead, use a fresh and floral gin, perhaps with citrusy grapefruit notes.

Unlike rum, throughout the evolution of mixology, gin has withstood the introduction of another important distillate, one destined to change history as well—vodka—on account of its characteristics described above.

Beginning in the 1970s, vodka has played a fundamental role in mixology and in American (and therefore global) drinking habits. In fact, the vodka market surpasses that of all other spirits, especially rum, which has been essentially distorted to support sales, becoming lighter, more tasteless.

While rum production has morphed drastically in order to maintain its status as the backbone of cocktails, the classic and traditional English gin, in particular London Dry, has remained almost unchanged. Still, in many recipes it can be cut with vodka, to lighten aromas without lessening alcohol content.

In replicating these recipes today, we certainly would not need to cut a London Dry with vodka, because from among the vast array of gins available on the market, we can certainly find a gin with an aromatic profile perfectly suited to our modern mixing needs.

The historical journey outlined thus far, one that has led gin to inclusion in the Tiki style, speaks to just how much the realm of gin mixing has changed and evolved, driven by historical needs, culture, and even for fashion and trends.

Today we can state without a doubt that gin has not simply been rediscovered as a distillate: its elevated techniques and aromatic profiles have fostered a world in which hundreds of labels are now, more than ever before in history.

We are living in a new Golden Age of cocktails, and once again it falls to bartenders to create and innovate within the world of gin mixology, to preserve the memory of cocktails and keep preparations and recipes such as those of Trader Vic alive, and ensure that this great potential does not go unexpressed, nor forgotten in the thousands of gin cocktail variations, often consumed merely to be fashionable.

Royal Hawaiian

TYPE	ALCOHOL CONTENT	TECHNIQUE	GARNISH
TIKI	15.6% abv	Shake&Strain	None required

RECIPE (12 cl)
1 ½ oz (4.5 cl) CAORUNN GIN
1 oz (3 cl) fresh pineapple juice
½ oz (1.5 cl) freshly squeezed lemon juice
¾ oz (2.25 cl) orgeat syrup

PREPARATION
Homemade orgeat syrup
Prepare around 2 cups of sweet almonds, including some bitter ones. Peel them (carefully, as the bitter almonds contain hydrogen cyanide) and soak in tap water for about 30 minutes. Drain the almonds and then smash and reduce to a pulp, or process them in a food processor. Transfer the almond pulp to a bowl. Add about 3 cups of water and let rest for 3–4 hours, being sure to stir the mix from time to time. Then filter with a sieve, cloth or gauze and carefully squeeze to extract all the liquid components. Discard the almonds. Sweeten the almond milk with a cup of sugar and stir until completely dissolved. Finally, add more sugar to taste and a few drops of zagara, or orange blossom.

METHOD
Fill the shaker with ice, almost to the brim, and wait until the metal has cooled. With a strainer, remove the excess water by turning the shaker upside down. Add the recipe ingredients, starting with the non-alcoholic ingredients to avoid compromising the ice and unnecessary dilution. Follow with the other ingredients.
Close the shaker tightly. Shake in your favorite style for at least 10-12 seconds, energetically! It is very important that the shaking is "explosive" so that the ingredients mix well, cool properly and emulsify. Otherwise you will achieve exactly the opposite effect.

NOTE
The name derives from one of the best known resorts in Waikiki, Hawaii, Royal Hawaiian, opened in the 1920s, and at that time known as "Princess Kaiulani".

Kamaaina

TYPE	ALCOHOL CONTENT	TECHNIQUE	GARNISH
TIKI	10.5% abv	Shake&Pour	Fresh mint leaves and an exotic fruit stick

RECIPE (15 cl)
2 oz (6 cl) Scortese Citrusy Bitters
½ oz (1.5 cl) freshly squeezed lemon juice
½ oz (1.5 cl) Lazzaroni Triplo (triple sec)
1 oz (3 cl) Coco Lopez
1 oz (3 cl) HOXTON GIN

METHOD
Prepare directly in a very cold shaker. Pour in all the ingredients except the Bitter Citrusy soda. After shaking for 10-12 seconds, pour the contents into a ceramic coconut mug, as is the traditional serving custom with this drink. Finish with the soda, gently mix and garnish.

NOTES
Kamaaina, from the Hawaiian *kama'aina*, meaning "son of the earth", is a word that describes residents of Hawaii, regardless of their racial background. Today it has taken on a meaning similar to "old-timer", in the sense of something "as it once was".

Pogo Stick

TYPE	ALCOHOL CONTENT	TECHNIQUE	GARNISH
TIKI	12.3% abv	Flesh Blend (Milkshake Mixer)	Fresh calamint

RECIPE (12+3 oz ice)
2 oz (6 cl) BEEFEATER 24
¼ oz (0.75 cl) liquid sugar syrup
½ oz (1.5 cl) freshly squeezed lime juice
¾ oz (2.25 cl) fresh pineapple juice
¾ oz (2.25 cl) fresh squeezed Yellow grapefruit juice
3 oz shaved ice

METHOD
Pour everything into an electric blender and process for a few seconds with precisely the indicated amount of ice. By doing so, once the contents are poured into the appropriate glass, the ice will remain on the surface and form a kind of "crust" (that will slowly dilute and keep the drink at temperature). Use a medium goblet or an old style champagne glass.

PREPARATION
Shaved ice is a "snowy" ice made with a dedicated machine called an ice shaver, or by grating it directly from a large cube of slow ice with an ice planer.

NOTES
According to the original recipe, this cocktail is garnished with a stick of rock candy, which is dipped in the drink to slightly sweeten it. Particularly suited to English palates of that time, the Pogo Stick is a unique drink of Trader Vic's, who had the habit of revising drinks at his various new openings around the world, introducing traditional spirits that would be particularly appreciated by new customers.

Fog Cutter

TYPE	ALCOHOL CONTENT	TECHNIQUE	GARNISH
TIKI	20% abv	Shake&Pour	Fresh mint leaves and an exotic fruit stick

RECIPE (24 cl)
2 oz (6 cl) Spanish style rum
1 oz (3 cl) Nardini Brandy
½ oz (1.5 cl) CITADELLE GIN
2 oz (6 cl) freshly squeezed lemon juice
1 oz (3 cl) freshly squeezed orange juice
½ oz (1.5 cl) orgeat syrup
Sweet sherry, floated

METHOD
Prepare directly in a very cold shaker. Pour in all the ingredients except the sherry. After shaking for 10-12 seconds, pour all the contents into a specially designed tiki mug, and finish with the correct amount of sweet sherry, floating style.

PREPARATION
Orgeat syrup
See page 159

NOTES
The Fog Cutter is a classic cocktail by Victor "Trader Vic" Bergeron, featured in his iconic 1947 book *Trader Vic's Bartender's Guide*. At Trader Vic's, this vintage drink was limited to two per person… In fact, in his bartending guide, it assessed as follows: "Fog cutter, hell. After two of these you won't even see the stuff".

Scorpion Kelbo's

TYPE	ALCOHOL CONTENT	TECHNIQUE	GARNISH
TIKI	18.5% abv	Shake&Pour	Calamint, passion fruit, orchid

RECIPE (14 cl + 8 oz crushed ice)
¾ oz (2.25 cl) freshly squeezed lime juice
¾ oz (2.25 cl) freshly squeezed orange juice
¾ oz (2.25 cl) Nardini Brandy
1 oz (3 cl) Jamaican Dark Rum
½ oz (1.5 cl) BICKENS London Dry Gin
½ oz (1.5 cl) passion fruit syrup
½ oz (1.5 cl) orgeat syrup
8 oz crushed ice

PREPARATION
Orgeat syrup
See page 159

Passion fruit syrup
Cut the passion fruit in half and use a spoon to empty the contents completely into a fine mesh strainer. With the back of the spoon, massage the pulp to obtain a juice. Weigh it, then add the same amount of white sugar. Stir until dissolved. Next, measure the amount obtained in milliliters and add the same amount of liquid sugar. Pour everything into a plastic bag, add the seeds (previously separated), vacuum-seal and let rest overnight. The following day, strain well. Your syrup is ready.

METHOD
This is Trader Vic's version of the hugely popular Scorpion from the famous Kelbos's restaurant in Los Angeles. Prepare directly in a very cold shaker. Pour in all the ingredients, including the crushed ice. After shaking for 10-12 seconds, pour all contents into a Scorpion Bowl. If the drink is being served according to its tradition—that is, for 3 people—multiply the recipe amounts as proportionally needed.

Chapter VII

100 NOT-TO-MISS GINS

"A sailor best working compass is a glass completely full of genever"
An old Dutch saying

According to an app that provides information on gins—nowadays there is an app for everything— over five thousand gins are produced around the world. An impressive number indeed, even to the most savvy of bartenders, and one that surely provides the clearest perspective on gin's remarkable global success: it really is the "spirit" of the planet. With production no longer limited to gin's natural or adopted homelands, namely Holland, England, and even the United States, countries as far-flung and diverse as Italy (with an ever-increasing number, in fact), Brazil, France, Japan, Croatia and New Zealand can now be counted among gin's top producers. And while juniper remains the universal "compass" guiding producers throughout their various creations, the bulk of all the other spices, roots and fruits used in making gin represents their playground of innovation and experimentation. This chapter outlines a selection of one hundred gins, each endorsed by Samuele Ambrosi, beginning with his ten favorite labels. All are gins that merit your attention, on account of their quality, and their originality. So get ready. It's time to fill those glasses!

BEEFEATER 24
45% ABV

STYLE London Dry Gin – Super Premium Gin

ORIGIN Kennington, London – United Kingdom

BOTANICALS 12. Juniper, lemon and sweet orange peel, grapefruit peel, coriander, iris, angelica root and seeds, licorice root, Sencha tea, Chinese green tea, bitter almond oil.

PRODUCTION:
After 18 months of research, conducted by someone I consider "the king of London gin", master distiller Desmond Payne, Beefeater 24 was launched on October 30, 2008. Distillation lasts 8 hours, to which the botanical maceration (steeped 24 hours in alcohol) is added.. Up to half a ton of botanicals are used for each production batch, and the technique is a multi-shot, which is cut with a neutral spirit.

For Payne, introducing tea into the recipe was a novel and challenging experience, given that the tannic part present therein emerges near the distillation tails. His solution was to cut the distillation in advance.

CURIOSITY:
• Beefeater was born in 1863, when James Burrough, a young chemist from Devonshire, bought the liquor firm Taylor & Son, located on Cale Street, Chelsea.

• Beefeater London Dry Gin quickly became the distillery's flagship product, a success that convinced Burrough to officially register the recipe in 1895.

• The Burrough family sold the company to Whitbread in 1987, who in turn sold it to Allied Domecq in 1991 (which joined the Pernod Ricard group in 2005).

• Beefeater 24 takes its name from the steeping time of botanicals, 24 hours.

LABELS:
• *Beefeater London Dry Gin*: 40% abv. With juniper, coriander, angelica, iris, lemon and orange peel, liquorice, cassia and nutmeg.

• *Beefeater's Burrought's Reserve Gin*: 43% abv. Produced for the first time in 2013, and made in a small historic alambic (243 liters) that dates to the 19th century.

• *Beefeater Crown Jewel Gin*: 50% abv. Produced from 1993 to 2009. The bottle bears the names of the Queen's 8 ravens that guard the Tower of London.

COCKTAILS:
White Lady, Negroni, Paradise, Limmer's Gin Punch, Bee's Knees, London Mule, Gimlet, Corpse Reviver No. 2, Bramble, Last Word, Seventh Heaven, Monkey Gland.

GARNISH GIN&TONIC:
Pink grapefruit peel, Garda lemon peel, fresh violettes.

RECOMMENDED TONICS:
Scortese Pure Tonic, Fever Tree Indian, Indi, A.Q. Monaco, 1724, Le Tribute

WEB www.beefeater24.com

BOMBAY SAPPHIRE DRY GIN
40% ABV

STYLE London Dry Gin

ORIGIN Laverstoke Mill, Hampshire – United Kingdom

BOTANICALS 10. Juniper and iris from Tuscany, lemon peel from Murcia, coriander seeds from Morocco, angelica root from Saxony, bitter almonds from Spain, licorice from China, cassia bark from Indonesia, cubeb berries from Java, grains of paradise from West Africa.

PRODUCTION:
Their secret lies in the placement of the basket, external to the rectification column where the botanicals are placed. Steam passes through the column to a copper basket, containing a basket that holds the recipe's botanicals. The steam extracts the aromatic oils, yet without damaging or compromising them.

HISTORY:
Thomas Dakin built his distillery in the North West of England in 1760. The Bombay Company, on the other hand, started up in 1957, when Allan Subin created a gin using an infusion basket. He later turned his attention to the Greenall distillery and, around 1960, launched Bombay Dry Gin. A turning point for this gin came along in 1987, Michel Roux created Bombay Sapphire, with its blue bottle inspired by the famous Star of Bombay sapphire.

CURIOSITY:
- The master herbalist is an Italian, Ivano Tonutti.

LABELS:
- *Bombay Dry Gin*: 40% abv. Produced with only 8 botanicals.
- *Bombay Sapphire East*: 42% abv. Launched in 2012 with Sapphire botanicals, with the addition of Thai lemongrass and Vietnamese black pepper.
- *Bombay Amber*: 47% abv. Launched in 2015. Features black cardamom, nutmeg and bitter orange, and aged for 6 weeks in French vermouth barrels.
- *Star of Bombay*: 47.5% abv. This gin takes its name from the 182-carat sapphire that actor Douglas Fairbanks gave to Mary Pickford. The botanical selection is characterized by ambrette seeds, bergamot and licorice root.
- *Bombay Sapphire English Estate*: 41% abv. Based on the Bombay Sapphire recipe, yet rebalanced with spearmint, toasted hazelnuts and rose hips.

COCKTAILS:
Bramble, Gin Fizz, White Lady, Gin Smash, Bee's Knees, Ramos Fizz, Gin Cobbler, London Mule, Cardinale, Last Word, Monkey Gland, Vesper Martini.

GARNISH GIN&TONIC:
Garda lemon peel, pink grapefruit peel, slice of liquorice bark.

RECOMMENDED TONICS:
Cortese Bio 1959, Fever Tree, Goldberg, AQ Monaco, Cortese Strong, Erasmus Bond

WEB www.bombaysapphire.com

Bulldog Gin
40% ABV

STYLE London Dry Gin – Ultra Premium Gin

ORIGIN Warrington, Cheshire – United Kingdom

BOTANICALS 12. Lemon peel, almonds, cassia, lavender, juniper, iris, liquorice, angelica from Germany, coriander, Dragon's eye (Dimocarpus Longan), white poppy, lotus leaves (Nymphaea).

PRODUCTION:
Anshuman Vohra has created one of the world's best-selling brands, by going in search of origins of gin. His research led him to the historic Greenall's Distillery, one of the oldest in the world (since 1761) that is still operational. Here he met master distiller Joanne Simcock Moore (one of the rare female master distillers in this business) and launched a new concept project of Premium Gin, in January, 2007. Bulldog was born of carefully selected maize from Norfolk, distilled 3 times, and lastly a fourth with the 12 botanicals in infusion, after having lowered the alcohol content from 96% to 76% abv. Infusion takes place in a historic copper pot still throughout the night, while distillation happens the following day. Lowering of the final alcohol content is done using water from Lake Vyrnwy, in Wales.

CURIOSITY:
- After raising his starting capital ($600,000) from among friends and private individuals, Vohra quit his job in July 2006 and launched Bulldog Gin 6 months later.

- Bulldog is released in a black bottle with purple hues. The brand is named after Sir Winston Churchill, nicknamed "the British bulldog" on account of his tenacity.

- Bulldog's relationship with Campari dates to 2014, when the Italian group approached Vohra to distribute it, thus closing the gap in the Premium Gin category (with a call option to acquire the brand by 2020).

- The bottles are made from eco-compatible and biodegradable materials, and the gin has been granted kosher certification.

- Apparently, in addition to paying homage to Churchill, the name is associated with the Chinese zodiac: the launch took place during the year of the dog, which also influenced the choice of name.

LABELS:
- *Bulldog Gin Extra Bold*: 47% abv. Produced for the first time in 2013.

COCKTAILS:
Bramble, Gin Fizz, White Lady, Gin Smash, Bee's Knees, Gin Spicy Fifty, Ramos Fizz, Gin Cobbler, Red Lion, Caricature Cocktail by Gary Regan.

GARNISH GIN&TONIC:
Garda lemon peel, fresh lavender, lychee.

RECOMMENDED TONICS:
Scortese Pure Tonic, Thomas Henry, Goldberg, Indi

WEB www.bulldoggin.com

CAORUNN GIN
41.8% ABV

STYLE Scottish Dry Gin

ORIGIN Cromdale, Scottish Highlands – United Kingdom

BOTANICALS 6 traditional: juniper, coriander, lemon peel, orange peel, angelica root, cassia bark; 5 Scottish: sorb fruit, erica, dandelion, bog myrtle (*Menyanthes trifoliata*), Coul Blush apples.

PRODUCTION:
Handcrafted in small batches of about 1,000 liters each at the Balmenach Distillery, where master distiller Simon Buley desired to return to his origins by using Celtic botanicals. The base grain spirit is triple-distilled.

The distiller is heated exclusively by steam, in a special still dating to 1920 located at the base, with a body measuring 91 cm in diameter. Inside, upon the 4 perforated trays in stainless steel, approximately 20 kg of botanicals, the amount called for in the recipe. The neutral alcohol present in Vat No. 2 is channeled towards Vaporiser No. 2, where the heat transforms the alcohol into steam, which then passes to the still's base, coming into contact with the botanical components. The noble parts of the recondensed steam are then re-channeled into Vat No. 1. Next, completely emptied, Vat No. 2 is refilled with the contents of Vat No. 1, then repeating the steps to create a concentrated Caorunn Gin. Lastly, the alcohol content is reduced using pure water from a source right next to the distillery.

CURIOSITY:
- The Balmenach Distillery's first license dates to 1824, opened by James McGregor. It is one of the oldest distilleries in all of Speyside, and the property belonged to the founders until 1922, when it was incorporated by a consortium, and then later was bought by DCL (Distillery Company Limited, now Diageo). It was closed from 1993 to 1998, then repurchased by its former owners.

- "Caorunn" (pronounced: ka-roon) is a Celtic word meaning "rowan berry".

- The red star (an asterisk, for some) on the bottle represents the 5 Scottish botanicals used.

COCKTAILS:
Bee's Knees, Apple Martini, Corpse Reviver N.2, Collins, Gin Toddy, White Lady, Gin Smash, May Fair Cocktail, Bramble, Angel Face, Straits Gin Sling, Maravel Sling.

GARNISH GIN&TONIC:
Thinly sliced apple, Garda lemon peel, fresh chamomile.

RECOMMENDED TONICS:
Scortese Pure Tonic, Fever Tree Indian, Thomas Henry, East Imperial

WEB www.caorunn.com

GINEPRAIO TUSCAN DRY GIN
45% ABV

STYLE London Dry Gin – Organic

ORIGIN Tuscany – Italy

BOTANICALS 9. Rose hips (rosa canina), helichrysum, angelica, lemon and orange peel, coriander and 3 different varieties of juniper (from Arezzo, Chianti and the Tuscan Maremma).

PRODUCTION:
Created by two friends, Enzo Brini and Fabio Mascaretti, whose enological approach has resulted in the perfect recipe for a Tuscan London Dry Gin, one in which each botanical must be of certified organic origin.

To distil the basic alcohol, produced 100% from soft wheat from Tuscany's Mugello region, they turned to the Sacchetto Distillery in Cuneo; while for redistillation of the refined botanicals, they chose the Deta Distillery located in Barberino in Val D'Elsa (Tuscany). The botanicals are infused separately in alcohol, for varying times that depend on the spices and the alcohol content. The infusions are then combined and distilled in a small, 300-liter pot still. Lastly, the distillate is left to rest for about ten days before bottling takes place.

CURIOSITY:
- Ginepraio is an organic Tuscan London Dry Gin, the only gin whose alcohol and locally-sourced plants are fully certified.

- The name derives from the Italian saying "cacciarsi in un ginepraio" (meaning something like "get oneself into a mess" or "tangled up"). The creators say, in fact, that they got into "a mess" while seeking out their all-certified, Tuscan and organic ingredients.

- Enzo Brini is not new to the world of organic production: with his family he co-owns a winery in Montepulciano, which has been certified organic since 1990.

- The quantity of juniper used for each single batch is significant: about 40 grams per liter.

LABELS:
- *Ginepraio Navy Strenght*: 57% abv. Produced for the first time in 2017 and subsequently aged in Opus signinum ("cocciopesto" in Italian) amphorae for about 6 months. Opus signinum is an ancient material, composed of stone aggregates, ground brick, water and hydraulic lime. It permits the distillate to micro-oxygenate, lending it superior olfactory complexity. The label bears the year of distillation.

COCKTAILS:
Martini, Negroni, Claridge, Bramble, Gin Fizz, White Lady, French 75, Bee's Knees, Red Snapper, Ramos Fizz, Red Lion, Suffering Bastard, British Spring Punch.

GARNISH GIN&TONIC:
Garda lemon peel, edible rose petals, pink grapefruit peel.

RECOMMENDED TONICS:
Cortese Classic and Cortese Light, Thomas Henry, Goldberg, Indi, Fever Indian, AQ classic

WEB www.levantespirits.it

MARTIN MILLER'S GIN
40% ABV

STYLE London Dry Gin

ORIGIN United Kingdom

BOTANICALS 10, including juniper, Florentine iris, cassia bark, licorice root, citrus peel (Seville orange, lemon, lime), cinnamon, coriander, nutmeg and angelica root.

PRODUCTION:
Produced at Langley Distillery in Langley Green near Birmingham, Martin Miller's is distilled in perfect London Dry Gin style in what is called "the Rolls Royce of pot still" (also known as "Angela"), a still designed by John Dore & Co in 1903. Production is characterized by the union of two distillates from different batches: one from the full recipe, while the other leaves out the citrus component, the latter of which takes place in a "steeping overnight still" (meaning, the botanicals are immersed in alcohol the evening before distillation).

CURIOSITY:
- The water used to reduce the alcohol content here is entirely unique. Called "the silent element", the water used comes from Borgarnes in Iceland, over 3,000 miles away. It is considered by many one of the world's purest waters, a purity determined by natural filtration through several meters of volcanic lava, as well as owing to glacial source, glaciers formed millions of years ago in an area of the world boasting some of the lowest pollution levels known.
- Launched in 1999 by Martin Miller, an antiques and real estate entrepreneur, owner of Miller's Residence in Notting Hill.
- Some say a "secret ingredient" lurks among the botanicals, which could be cucumber (according to some sources), which is not as a "flavoring" but rather as a "dry agent" in the finish.

LABELS:
- *Martin Miller's Westbourne Strength Gin:* 45.2% abv. Characterized by a different balance between the two distillates.
- *9 Moons Aged Gin:* 40% abv. Each barrel produces a batch equal to about 2,000 bottles for circulation. The barrels are made from American oak that previously contained bourbon. Aging takes place for 9 months in Borgarnes (Iceland), where, given the extremely low temps, the process is particularly slow.

COCKTAILS:
Martini, Corpse Reviver N.2, Negroni, John Collins, Clover Club, Claridge, Bennett Cocktail, British Spring Punch, London Mule, Bijou, Pegu Club, Gin Fix.

GARNISH GIN&TONIC:
Garda lemon peel, cucumber, lime.

RECOMMENDED TONICS:
Fever Tree, Goldberg, Cortese Bio Originale, Franklin Indian T, AQ Monaco

WEB www.martinmillersgin.com

PANAREA ISLAND GIN
44% ABV

STYLE Distilled Compound – Premium Gin

ORIGIN Turin – Italy

BOTANICALS 7. Alpine and Tuscan juniper, cardamom, lemon and orange peels from Sicily, coriander, very rare Panarea myrtle, Panarea thyme.

PRODUCTION:
Panarea was launched at the end of 2015, the result of meticulous and costly research into botanicals that originate on the island from which this gin takes its name. To make the most of its properties, each botanical is worked individually. First is the selection of the neutral base alcohol, distilled twice, made from different grains of Italian origin. Next comes the individual macerations in neutral alcohol, each a carefully selected botanical. These will have different alcoholic concentrations, and different timings based on the characteristics of the single botanical. Each macerate is then filtered and the extract is carefully vacuum distilled in a small batch still made entirely of steel. Thus individual spirits of about 70-73% abv are obtained, and are then meticulously combined to create the perfect balance of Panarea Island Gin.

CURIOSITY:
• The Inga family's long tradition and experience in the production of high quality liqueurs and spirits dates to 1832 in Noto, Sicily, tradition and experience that materialized in 1930 with the birth of the first factory in Piedmont, opened Gaetano Inga, who decided to relaunch the ancient Amaro Gambarotta. In the 1970s, however, grappa production began, with the well-known Libarna brand. Today, this incredibly rich family tradition is continued thanks to brothers Federico and Lorenzo Inga.

• The name "Panarea" is a tribute to both the family's Sicilian origins and their rich cultural heritage, as well as the same-named island in the Aeolians, where many of the botanical essences that lend distinctive complexity to this gin grow.

LABELS:
• *Panarea Sunset Gin*: 44% abv. The base used comes from Island itself, in different proportions of course, to which individual distillates of Panarea basil and yellow grapefruit are added.

COCKTAILS:
Bramble, Gin Fizz, Gimlet, White Lady, French 75, Bee's Knees, Ramos Fizz, Caricature Cocktail by Gary Regan, Clover Club, Singapore Sling, Monkey Gland.

GARNISH GIN&TONIC:
Lemon peel, fresh rosemary, basil leaves.

RECOMMENDED TONICS:
Scortese Pure Tonic, Thomas Henry, Goldberg, Indi, Fever Tree Mediterranean, AQ Herbal

WEB www.panareagin.it

Rivo Gin
43% ABV

STYLE Distilled Compound – Premium Gin

ORIGIN Lake Como – Italy

BOTANICALS 12, including juniper, creeping thyme, lemon balm, winter savory, pimpinella, coriander, cardamom, angelica.

PRODUCTION:
Rivo Gin is distilled at the Antica Quaglia Distillery. Most of the botanicals contained in Rivo gin are collected at specific times of the year, and each is distilled separately to extract the best component.

The botanicals are left to macerate for about 10 days, in 70% abv Italian wheat alcohol, while the harvested botanicals themselves do so for only 48 hours. Each botanical is then smashed and the extract is distilled individually in a double-bottomed pot still. Measurements and times in each distillation differ from all the others, to preserve the quality of every single botanical component. Subsequently, the various distillates are combined in various proportions, then bottled after resting for a few weeks at 43% abv.

CURIOSITY:
- This project was born in 2014 from an idea of Marco Rivolta and his mother Gianna, who, as a hobby, collects and grows herbs on Lake Como (a terroir offering remarkable botanical diversity) to make this very special gin.

- The project lasted nearly two years, having started with 50 different botanicals, and relying on the advice of Samuele Ambrosi.

- The supporting structure is a very classic London Dry Gin with juniper in the dominant role, yet sustained and well-integrated with coriander, cardamom and angelica. The individual botanicals are collected by a group of expert foragers, three times a year.

- In just one year, the Rivo has become the most awarded Italian gin in the world.

LABELS:
- *Rivo Sloe Gin*: 30% abv. Wild Italian blackthorn berries are hand-picked when the first cold breaks the skin, thus intensifying the aromas. The berries are then left to rest in the Rivo Gin, which is then filtered and bottled. Rivo's annual production is only one thousand bottles.

COCKTAILS:
Bramble, Gin Fizz, White Lady, Gin Smash, Gin Julep, French 75, Bee's Knees, Ramos Fizz, Gin Cobbler, Red Lion, Last Word.

GARNISH GIN&TONIC:
Garda lemon peel, orange peel, fresh lemon balm.

RECOMMENDED TONICS:
Scortese Pure Tonic, Thomas Henry, Goldberg, Indi, Fever Tree Indian

WEB www.rivogin.com

ROBY MARTON GIN
47% ABV

STYLE Distilled + Cold Compound – Premium Gin

ORIGIN Treviso – Italy

BOTANICALS 11, including citrus peel, cinnamon, licorice root, aniseed, juniper berries, horseradish root, ginger, cloves, pimento and cardamom.

PRODUCTION:
Produced at Distilleria Montegrappa, in the Vicenza province in Italy.

The production takes place in two distinct phases.

– In the first, a distillate is made with a base spirit and juniper, thus obtaining the gin's base. The alembic is a small, old copper pot still.

– In the second phase, for about 10% of the total volume, a cold infusion is made (cold compound), where all the botanicals except the juniper are macerated for about twenty days in 50-liter drums, into which the previously produced neutral alcohol, triple-distilled, is poured. Something not well known to many, yet revealed by Roberto Marton himself for this book, is that the recipe varies slightly from batch to batch, with the aromatic characteristics of the botanicals meticulously evaluated (characteristics that naturally are not always uniform). Eventually the gin undergoes a coarse microfiltration, noticeable by observing the distillate's slightly cloudy appearance—a quality Marton wanted to preserve, to maintain some elements precious to his gin's taste notes. It is bottled after a few days of rest.

CURIOSITY:
- Roberto Marton had started thinking about a gin before the real potential of this distillate in Italy was known. A trailblazer, Marton also wanted to "play" with the most difficult challenge—creating a product that he could identify with, with a style as complex as his own and, to use his own words, somewhat of a difficult customer.

LABELS:
- *Big Gino*: 40% abv. Differs only in the number of botanicals (juniper, pomelo peel, pimento berries) and in the infusion times.
- *Tonka Gin*: 47% abv. The base used is the same as the original Roby Marton, with a spirit redistilled with only juniper. Subsequently, Venezuelan Tonka beans are left to macerate.

COCKTAILS:
Gin Smash, French 75, Bee's Knees, Ramos Fizz, Gin Julep, Gin Cobbler, Hanky Panky, Martinez, Italian Mule (created by Roberto Marton), Monkey Gland.

GARNISH GIN&TONIC:
Garda lemon peel, slice of ginger, slice of liquorice.

RECOMMENDED TONICS:
Scortese Pure Tonic, Ledger's, Thomas Henry, Goldberg, Indi, Fentimans

WEB www.robymarton.com

TANQUERAY TEN GIN
47.3% ABV

STYLE Distilled Compound – Ultra Premium Gin

ORIGIN Leven, Fife – United Kingdom

BOTANICALS 8. Juniper, coriander, angelica, liquorice, lime, grapefruit, orange, chamomile.

PRODUCTION:
A wheat spirit produced at the Cameronbridge Distillery, using the same base as Sterling Vodka. With Tanqueray Ten, we are talking about a gin distilled 4 times. The heart of Tanqueray London Dry Gin is redistilled in a small copper pot called "Tiny Ten" in which chamomile, grapefruit, oranges and lime will be poured—fresh, not dried. The result is then blended with Tanqueray London Dry Gin to be distilled again. The botanicals are immersed in alcohol and distilled immediately.

HISTORY:
The original distillery was founded by Charles Tanqueray in 1830 in the Bloomsbury district. In 1898, Tanqueray merged with Gordon & Company, while in 1986, Tanqueray & Gordon Co. was acquired by United Distillers (now Diageo).

The distillery was destroyed by German bombs during the Second World War, and the only alembic that managed to survive was "Old Tom". Work began on the new distillery in 1951. It was transferred to Essex in 1989 and, after just 10 years, moved back to its current location.

CURIOSITY:
• Tanqueray Ten was presented in 2000. The bottle was redesigned 4 years later by the London agency Design Bridge, who chose to bestow upon it a more Art Deco style.

• During Prohibition, when the illegal export of gin from England continued, it would not have been unlikely to find Tanqueray on the black market. Apparently, bottles were thrown directly from the ships into the sea, near the coast, in waterproof barrels.

LABELS:
• *Tanqueray London Dry Gin*: 47.3%– 43.1% abv. With juniper, angelica, coriander and liquorice.

• *Tanqueray Rangpur*: 41.3% abv. A 2006 release that features Bengali Rangpur lime.

• *Tanqueray Old Tom*: 47.3% abv. Launched in 2014, a reinterpretation of the original recipe, a pre-1920 Old Tom. Sweetened with beet sugar.

• *Tanqueray Flor de Sevilla*: 41.3% abv. Launched in 2018, with bittersweet Seville oranges and orange blossoms.

• *Tanqueray Bloomsbury*: 47.1% abv. Made by master distiller Tom Nichol, who took inspiration from a recipe by Charles Waugh Tanqueray, son of the founder, when the distillery was still based in London's Bloomsbury district. Launched in 2015 with a limited edition of only 100,000 bottles, this gin uses Italian juniper, winter savory, coriander, angelica and cassia.

• *Tanqueray Malacca*: 41.3% abv. Presented in 1990 and conceived as an Old Tom Gin, this one was withdrawn from the market in 2000, and later re-launched in a limited edition on February 4, 2013. Inspired by a recipe of Charles Tanqueray's of 1839 (proposed as "non dry gin"), this gin, in addition to its classic botanicals, is rumored to include grapefruit, licorice, lavender and jasmine…

• *Tanqueray Lovage*: 47.3% abv. Born with the help of Joanne McKerchar, historical researcher who oversees the Diageo archives, and the well-known bartender Jason Crowley. Launched in 2018, its recipe (from 1839) calls for juniper, angelica, coriander, celery seeds and roots, nettle, cinchona bark, Java pepper, chamomile and winter savory. It is named for its primary botanical component, lovage.

COCKTAILS:
Martini Cocktail, Negroni, John Collins, Clover Club, British Spring Punch, London Mule, Bijou, Mayfair Cocktail, Pegu Club, Gin Fix, White Lady, Paradise, Gimlet.

GARNISH GIN&TONIC:
Grapefruit peel, lime peel, fresh chamomile flowers.

RECOMMENDED TONICS:
Goldberg, Franklin Indian T, AQ Monaco, Scortese Pure Tonic, Fever Tree Indian, Thomas Henry, Indi

WEB www.tanqueray.com

AVIATION AMERICAN GIN
42% ABV

WEB www.aviationgin.com

STYLE American Dry Gin

ORIGIN Portland, Oregon – United States

BOTANICALS 7. Juniper, coriander, lavender, Indian sarsaparilla, sweet orange peel, coriander seeds, anise.

PRODUCTION Created and developed by Christian Krogstad and Ryan Magarian in 2006, this gin is produced at the House Spirits Distillery, where all 7 botanicals are immersed in grain spirit for 48 hours, then redistilled in a specifically built pot still style alembic (about 1,500 liters). Alcohol content is later lowered with water from Cascade Mountain. The master distiller is Christian Krogstad.

NOTES House Spirits Distillery sold the brand to New York-based Davos Brands LLC in 2016, while continuing to distil the product.

In February 2018, actor Ryan Reynolds acquired some holdings of the Davos brand.

In this American Dry Gin style, the juniper component is deliberately reduced to allow space for the other botanicals, different from classic London Dry Gin.

G&T GARNISH Garda lemon peel, lavender flowers.

COCKTAILS South Side, French 75, Floral Gin Fizz.

BATHTUB GIN
43.3% ABV

WEB www.ableforths.com/bathtub-gin/

STYLE Cold Compound Gin

ORIGIN Tunbridge Wells – England

BOTANICALS 6 stated. Juniper, coriander, cloves, citrus peel, cinnamon, cardamom (also ginger, according to some).

PRODUCTION Produced at Master of Malt in Tunbridge Wells, where the neutral distillate is infused with botanicals for variable times, and continuously monitored via periodic analysis of samples to provide uniformity and maximum quality to the consumer. It is then filtered and bottled.

NOTES Founded in 2011, Ableforth's was created by 3 friends who simply desired to produce a delicious range of spirits.

Initially launched with productions of 30-60 bottles at a time.

Professor Ampleforth, also known as "The Ableforth Series" is a line of products owned by Maverick Drinks.

The packaging is altogether winning, a truly unique style that recalls an era when a bottle's contents were not visible from the outside: it is wrapped in brown paper ("parcel" style) and features handmade writings and is closed with cord and black sealing wax.

RELEASE Bathtub Gin Cask Aged: 43.3% abv. Aged from 3 to 6 months in ex-whisky American barrels.

Bathtub Gin Navy Strength: 57% abv. This gin won Best Compound in the world at the 2017 International Wine & Spirit Competition.

Bathtub Gin Navy Strength Cask-Aged: 57% abv. Aged at least 6 months in ex-whisky American barrels.

Bathtub Gin Old Tom Gin: 42.4% abv.

Bathtub Gin Sloe Gin: 33.8% abv. Made with 250 gr of plums.

G&T GARNISH lemon zest, thin slice of ginger.

COCKTAILS Hanky Panky, Martinez, Gin Old Fashioned.

BERKELEY SQUARE GIN
40% ABV

WEB www.berkeleysquaregin.com

STYLE London Dry Gin

ORIGIN Warrington, Cheshire – England

BOTANICALS 8. Juniper, angelica, basil, coriander, Java pepper, Makrut lime leaves, lavender, sage.

PRODUCTION A product launched in 2008 by Greenall's Distillery (Quintessential Brands Group) and created by the amazing master distiller Joanne Simcock Moore, seventh master distiller in the distillery's 250-year history. One of the world's few female master distillers, she was inspired by herbs that grow "in a timeless English garden", favoring some fresh Mediterranean botanicals such as basil and sage. The production method is described as "two-day bouquet-garni process" in a traditional, small batch copper pot still, filled with a triple-distilled neutral wheat alcohol. The heart of the botanicals, including the Makrut lime, is left to macerate directly for 24 hours. The following day it's the turn of the other botanicals (sage, entirely harvested by hand, basil, lavender), which are wrapped in muslin and left to macerate all together with the previous botanicals for a day. At the end of these 48 hours of steeping, the fourth—slow—distillation at 80 ° C happens, designed to capture only the noblest and most refined essences from this precious distillate.

NOTES The name derives from same-named square in London's West End, City of Westminster (currently the most exclusive region in England, seeing skyrocketing real estate prices). The name of the square, in turn, derives from the Berkeley family, namely John, third Baron Berkeley of Stratton, who, in spite of his wishes, was obliged to move here from Piccadilly.

The still used is No. 8, also known as "Baby".

The recipe chosen from among the various tests carried out was No. 5.

Basil is known as "The King of the Herbs".

G&T GARNISH Amalfi lemon peel, basil leaf, tuft of lavender.

COCKTAILS Martini Cocktail, South Side, Gimlet.

BICKENS LONDON DRY GIN
40% ABV

WEB www.camparigroup.com, www.bickensgin.com

STYLE London Dry Gin

ORIGIN Italy (also produced in the UK)

BOTANICALS 10. Juniper from the Balkans, cinnamon from Madagascar, orris root from Italy, coriander from Eastern Europe, nutmeg from the West Indies, lemon and orange peel from Spain, angelica root from France and Belgium, cassia from China, licorice from the Mediterranean.

PRODUCTION Produced at Langley Distillery (Alcohols Limited) in Langley Green near Birmingham, where it is distilled in perfect London Dry Gin style in what is called the "Rolls Royce of pot still", or "Angela"—a 3,000-liter still designed in 1903 by John Dore & Co. And here comes the stroke of genius: Bickens is partly distilled in Angela (No. 1) and partly in Jenny (No. 7), another 10,000-liter copper still made in 1995, which explains the writing on the bottle of their Premium Blended Gin: two spirits, from recipes of different eras and stills of different shapes, are perfectly combined.

G&T GARNISH Amalfi lemon peel, pink grapefruit peel, licorice stick.

COCKTAILS Singapore Sling, Dead Bastard, Marvel Sling, Negroni.

BIG GIN
47% ABV

WEB www.captivespirits.com

STYLE American Dry Gin

ORIGIN Seattle, Washington – U.S.

BOTANICALS 9. Cardamom, juniper, coriander, grains of Paradise, angelica, cassia, bitter orange peel, orris root, Tasmanian pepperberry.

PRODUCTION Distilled at Captive Spirits Distillery in Ballard, near Seattle, and produced from a 100% grain base spirit. Distilled in two 378-liter direct fire steel and copper pot stills, made by Vendome & Brass of Louisville (and named "Phyllis" and "Jean", after their grandmothers). Distillation takes place in perfect London Dry Gin style.

NOTES First created in March 2012. The distillery was launched 4 years prior by Ben Capdevielle and his girlfriend Holly Robertson, who were joined by a third partner, Todd Leabman, a builder who manages the company from a bureaucratic / administrative perspective.

The name is Capdevielle's dedication to his father, nicknamed "Big Jim", who was a Wisconsin craft distiller.

G&T GARNISH Sorrento lemons peel, orange peel, sprig of lavender.

COCKTAILS Tom Collins, Ramos Gin Fizz, Negroni.

BLACKWOODS VINTAGE DRY GIN
40% ABV

WEB www.blackwoodsgin.co.uk

STYLE London Dry Gin

ORIGIN Shetland – Scotland

BOTANICALS 19 stated. Juniper, angelica, sea thrift, lemon and orange peel, cassia, coriander, turmeric, violet blossom, iris, wild mint, nutmeg, mead wort/meadowsweet, licorice powder, ginger, thrift, wild thyme, euphrasia/eyebright, primrose.

PRODUCTION The distillery was founded in July 2002, while their gin was born in 2003. This product, however, is not made in Shetland, but rather by the Palmer group in the Midlands. The brand is managed by Blavod Wines & Spirits. Blackwoods Vintage Dry Gin is distilled in a John Dore & Co style small copper still.

NOTES The Blackwoods distillery was originally opened to produce whiskey, but after a very short time it was placed under an administrative seizure and repurchased by Catfirth Ltd, in Shetland. It then resumed working at full capacity, in an entirely different direction: gin.

The recipe speaks to Viking culture and reflects the dominating presence of the Vikings, who invaded Scotland between the 8th and 9th centuries.

The term vintage is applied because, as gin is made exclusively from local products yet the Shetlands cannot guarantee constant production of these products, the recipe varies from year to year based on Mother Nature's offerings.

Everything is hand-picked by the local crofters, bush by bush.

RELEASE Limited edition Vintage Dry Gin: 60% abv.

Botanical Vodka: 40% abv. Distilled 5 times.

G&T GARNISH Amalfi lemon peel, thin slice of ginger, thyme sprig.

COCKTAILS Southside, French 75, John Collins.

GREENALL'S BLOOM GIN
40% ABV

WEB www.bloomgin.com

STYLE London Dry Gin

ORIGIN Warrington, Cheshire – England

BOTANICALS 7. Juniper, Roman chamomile, honeysuckle, pomelo (also called Chinese grapefruit), angelica, coriander, Java pepper.

PRODUCTION Launched in 2007 and produced by Greenall's Distillery (Quintessential Brands Group), it was created by the amazing master distiller Joanne Simcock Moore, one of the world's few female master distillers and seventh master distiller in the distillery's 250-year history. Greenall's Bloom Gin is produced in a traditional small batch copper pot still, filled with a neutral double-distilled wheat alcohol. The botanicals are left to macerate directly for 24 hours before reactivating the third distillation.

NOTES Joanna talks about her inspiration: "I draw inspiration from my daily life—relaxing with chamomile tea and envisaging an English country garden".

RELEASE Bloom Jasmine & Rose Gin: 40% abv. Pink Distilled Gin with added jasmine tea and rose.

Bloom Strawberry Gin Liqueur: 25% abv.

Bloom Lemon & Edelflower Gin Liqueur: 25% abv.

Bloom Sloe Gin Liqueur: 28% abv. Limited edition born in November 2013 and produced by macerating hand-picked blackthorn berries.

G&T GARNISH Garda lemon peel, chamomile flower, pomelo, fresh cherry.

COCKTAILS Seventh Heaven, Corpse Reviver No. 2, Vesper Martini.

REISETBAUER BLUE GIN VINTAGE
43% ABV

WEB www.bluegin.cc

STYLE Austrian Dry Gin

ORIGIN Austria

BOTANICALS 27, including juniper, turmeric, lemon peel, angelica, coriander, licorice root, white pepper, ginger and orange peel.

PRODUCTION Manufactured since 2006 by Reisetbauer Qualitätsbrand GmbH, its base distillate is made from pure Mulan wheat grown in the upper Austrian countryside. Distilled in a small copper pot still, double-distillation, obtaining a high quality, very delicate neutral result. The acquavite produced, here called Feinbrand, is 84% abv and is macerated with botanicals for 3-4 days. It will then be redistilled, again in a pot still, to reach an alcohol content of 85% abv. It is then diluted with pure mountain water from northern Austria, bringing the alcoholic strength to that printed on the label.

RELEASE Sloe Berry Blue Gin: 28% abv.

G&T GARNISH Citrus peel, sliced ginger, red berries.

COCKTAILS Bramble, Gin Smash, London Mule.

BLUE RIBBON ESSENTIAL GIN
40% ABV

WEB www.spiritsland.com

STYLE London Dry Gin

ORIGIN France

BOTANICALS 14. Juniper, allspice, thyme, coriander, iris, ginger, cinnamon, cubeb berries, cardamom, angelica, cassia, lemon peel, orange, lime.

PRODUCTION Produced at the Distillerie des Terres Rouges, located in the small village of Coulonges la Rouge in central France. Made from a 100% French wheat brandy, distilled 5 times in a bain-marie in small stills, where the botanicals are first macerated individually in alcohol, with juniper, and then distilled separately in a Cognac style still. Subsequently, each distillate is combined with the others in proper proportions. In other words, several small batches of London Dry Gin are produced, which are eventually combined.

NOTES The Blue Ribbon was an award (in the form of a blue cross) that bestowed exclusively on the fastest, most intrepid navigators: namely, those who crossed the Atlantic Ocean from both America and Europe.

The recognition originates with the Knights of the Order of the Garter, founded in 1348 by Edward III of England.

The brand was launched in 2007 as one of the first Premium Gins on the market.

G&T GARNISH Amalfi lemon peel, sprig of fresh thyme, lime peel.

COCKTAILS Bronx, Vesper, Gin Sling.

BLUECOAT GIN
47% ABV

WEB www.bluecoatgin.com

STYLE American Dry Gin

ORIGIN Philadelphia, Pennsylvania – U.S.

BOTANICALS The list is completely secret. The only known botanicals are juniper berries, citrus peels, coriander, angelica and chamomile.

PRODUCTION Produced by Philadelphia Distilling (founded in 2005), this gin was launched in May 2006. It is entirely organic, starting with the juniper that hails from Italy. Distilled 5 times in a traditional pot still, designed and constructed entirely by hand. The master distiller is Robert J Cassell. Its discontinuous distillation lasts 10 hours; and, if I may say so, in perfect London Dry Gin style.

NOTES On the name: Bluecoats were the uniforms worn by armed militias during the American Revolution. This gin is, therefore, dedicated to the American spirit of independence and rebellion.

RELEASE Bluecoat Barrel Reserve Gin: 47% abv. Aged in American oak barrels from 3 to 6 months.

Bluecoat Edelflower Gin: 47% abv.

G&T GARNISH Fresh citrus peel, fresh chamomile, pink grapefruit peel.

COCKTAILS Bee's Knees, Gin Fizz, Satan's Whiskers.

BOBBYS DRY GIN
42% ABV

WEB www.bobbysdrygin.com

STYLE Schiedam Dry Gin (Distilled Compound)

ORIGIN Schiedam – Holland

BOTANICALS 8 stated. Juniper, rose hip, cubeb berries, lemongrass, coriander, cloves, cinnamon, fennel seeds.

PRODUCTION The masterpiece belongs to one Sebastiaan van Bokkel, who summarizes his creation as a successful mix of "Dutch courage" and "Indonesian spice". Launched in 2014, it is produced at the Herman Jansen Distillery, founded in 1777, and is thus one of the oldest distilleries in the world. Research on the mostly organic botanicals, each of which is distilled separately in 600- or 3,000-liter copper pot stills then expertly combined with the others, has been remarkably diligent.

NOTES This story begins in the early 1950s, when Jacobus Alfons ("Bobby" to family and friends), born in Naku (Ambon, Indonesia), emigrated to Holland, settling in Schiedam. Bobby loved genever, but he also craved the flavors of home, which led him to create a sort of home brewed gin. At his grandmother's house, van Bokkel found an old bottle of this gin, belonging to his grandfather and, fascinated by its story, he began to reproduce it.

Two years of hard work later, he arrived at the final recipe.

The bottle is in the classic genever bottle style, called a "kruik", but in glass instead of terracotta.

The silkscreen printing is artistically decorated with an Ikat pattern, a traditional Indonesian motif, lending the bottle a truly unique design.

RELEASE Bobby's Schiedam Jenever: 38% abv. Flavored with juniper, ginger, lemongrass, cardamom and cubeb berries.

G&T GARNISH Small piece of lemon grass, citron peel, chamomile flowers.

COCKTAILS Gin Julep, Last Word, Gin Sling.

BOOTLEGGER 21 GIN
47% ABV

WEB www.prohibitiondistillery.com

STYLE London Dry Gin

ORIGIN Roscoe, New York – U.S.

BOTANICALS 5. Juniper, coriander seeds, lemon verbena, orris root or iris, bitter orange peel.

PRODUCTION Produced since 1929 at the Prohibition Distillery, just outside New York in what was once an old fire station. The botanicals are steeped in 60% abv, 100% corn alcohol for 24 hours, then redistilled in a pot still of over 1,300 liters. The distillate is not filtered afterwards: the botanicals, placed in maceration at different times according to their specific characteristics, are removed after maceration and before distillation.

NOTES Inspired by the Prohibition era, a period recalled also on the product's label, which resembles a medical prescription: at that time, alcohol was only available by prescription.

In those years, however, American presidents always made sure that the White House never lacked an abundant supply of alcohol. The suppliers of the time, known as bootleggers, rum runners and moonshiners, risked everything to whet the thirst of a nation that had suddenly turned "dry". This period is also known as the Jazz Age and the Roaring Twenties.

The number 21 refers to the Twenty-first Amendment to the Constitution of the United States—the repeal of Prohibition.

The producers' goal was to work with a London Dry style of gin, and it has always stayed true to this.

RELEASE Bootlegger Barrel Aged Gin: 46% abv. Aged in barrels nearly 20 liters capacity for 12 months.

G&T GARNISH Lemon verbena, lemon peel, orange peel.

COCKTAILS Gary Regan's Caricature Cocktail, Dying Bastard, Monkey Gland.

BOXER GIN
45% ABV

WEB www.boxergin.com

STYLE Distilled Gin

ORIGIN United Kingdom

BOTANICALS 12 stated. Juniper (Nepalese and Bulgarian), angelica root, bergamot, coriander, lemon and sweet orange peel, orris root, licorice root, nutmeg, cassia bark, cinnamon.

PRODUCTION Produced on behalf of Green Box Drinks at Langley Distillery (Alcohols Limited) in Langley Green near Birmingham, where it is distilled in the "Rolls Royce of pot stills", also called "Angela", a still designed in 1903 by John Dore & Co. The base distillate is made from wheat from eastern England. Ten of the twelve stated botanicals are macerated in alcohol for 8 hours, then distilled. Two additional spirits are added to this already complete, if you like, London Dry Gin. This is an unusual procedure, of course, but guaranteeing the result is one of the most prepared makers in the gin field. Two types of juniper are used: the first is Bulgarian, distilled together in the LDG process, while the second is Himalayan, distilled at the source in Nepal to maintain a freshness unique to its kind. A similar procedure is used for the precious bergamot peels, whose essential oils, obtained by squeezing the peels directly, are then distilled separately.

NOTES The name refers to the most beloved sport (from a colonial perspective).

In addition to the classic sealed bottle, an 8.4 liter bag-in-box is available, to be able to refill the bottle and reduce product costs, resulting in economic and ecological benefits as well as benefits related to managing the production space. And the data to support this? Packaging has been reduced by 95%, transport weight by 45% and volume by 63%.

This gin is dedicated to Tom King, the first world heavyweight champion, who also became famous for his hand-to-hand fighting technique.

RELEASE Boxer: 40% abv. Produced only for the British market.

G&T GARNISH Bergamot peel, lemon peel, cucumber peel.

COCKTAILS British Spring Punch, Negroni, London Mule.

BRECON SPECIAL RESERVE GIN
40% ABV

WEB www.welsh-whisky.co.uk

STYLE London Dry Gin

ORIGIN Penderyn, Wales – United Kingdom

BOTANICALS 10 stated. Juniper, angelica, cassia, cinnamon, coriander, lemon and orange peel, licorice, nutmeg, iris.

PRODUCTION Produced by Penderyn Whiskey Distillery (The Welsh Whiskey Company) and launched in 2001 according to an over 100-year-old recipe. Distilled in small batches using a Welsh spirit produced entirely in the distillery, in a traditional copper pot still designed by Michael Faraday.

NOTES The story behind this tiny Welsh distillery: in the late 1990s, a group of friends gathered in a pub, dreaming of making a pure, precious Welsh Gold whisky like Penderyn's Gold Seam.

A gin born in the historic village of Penderyn on the southern tip of the Brecon Beacons, a location remarkable also for the quality of the water flowing nearby.

Today is the only Welsh gin around.

Two Brecon gins produced in Penderyn: the Brecon Special Reserve Gin and the Brecon Botanicals Gin. Both are prepared according to a well-defined recipe, then shipped to Wales to be blended with Penderyn Malted Barley Spirit and the alcohol content lowered with spring water from the National Park.

While the master distiller is Gillian Macdonald, a name that guarantees quality in the whiskey world, the base for this gin is made outside the distillery.

RELEASE Brecon Botanicals Gin: 43% abv. They use only 8 botanicals: juniper, lemon peel, orange and bergamot, coriander, cinnamon, cloves and saffron.

G&T GARNISH Lemon peel, liquorice root, fresh mint.

COCKTAILS Bebbo Cocktail, Corpse Reviver No. 2, London Mule.

BROCKMANS PREMIUM GIN
40% ABV

WEB www.brockmansgin.com

STYLE Cold Compound

ORIGIN United Kingdom

BOTANICALS 10 stated. Juniper and iris from Tuscany, angelica from Belgium and Saxony, coriander from Bulgaria, cassia bark from Indochina, lemon and orange peel from Murcia in Spain, licorice from China, almonds from Spain, blueberries and blackberries from various parts of Northern Europe.

PRODUCTION The distillate base is made from a certified English wheat distilled 4 times, in which all the botanicals are macerated (in my personal opinion some are essences, although excellent essences). The product is then filtered and bottled.

NOTES Brockmans is an independent English gin company created by 4 founding partners: Neil Everitt, Bob Fowkes and two of their friends. With over 60 years of experience in the beverage industry and extensive knowledge of gin making, they are writing a new chapter in the history of the English spirit.

Launched in 2010.

Apparently distilled by the G&J Greenall Group Ltd.

G&T GARNISH Lemon peel, blueberries, grapefruit peel.

COCKTAILS Florodora, Bramble, Gin Smash.

BROOKLYN HANDCRAFTED GIN
40% ABV

WEB www.brooklyngin.com

STYLE American Gin (Distilled Compound)

ORIGIN New York – U.S.

BOTANICALS 10. Added to the 5 citrus fruits (key lime, lemon, orange, lime, kumquat) are juniper, angelica, coriander, lavender and orris root.

PRODUCTION Born from an idea of Emil Jättne and Joe Santos, marketing expert and lead figure in some prestigious professional collaborations within the spirits sector. The distillate base is made from maize, but the peculiarity here is that all the citrus fruits are peeled by hand, one by one, while the juniper is crushed, also manually, before each batch. Produced in small batches using a copper pot still manufactured by Christian Carl. Distillation takes place at Warwick, in the Hudson Valley region of New York, not in the city proper.

NOTES The Brooklyn Distilling Company was founded in 1895, with a goal of producing alcohol from molasses and waste products generated by the sugar industry, but was closed in the early 20th century. Recently reopened, the distillery now produces Brooklyn Gin—practically its first distillate in the last 100 years. The name is an homage to a slice of American spirits history, despite the gin's being produced at the Warwick Valley Winery & Distillery.

The bottle is in perfect Art Deco style.

Released on the market in 2010.

For each batch, only 300 bottles are produced, taking 3 days.

G&T GARNISH Lavender sprig, fresh lemon peel, kumquat.

COCKTAILS Bramble, Florodora, French 75.

BULLARDS NORWICH DRY GIN
42.5% ABV

WEB www.bullardsspirits.co.uk

STYLE London Dry Gin

ORIGIN Norwich – United Kingdom

BOTANICALS 10. Juniper, Tonka beans, orange peel, lemon peel, cassia, licorice, angelica root, black pepper, cardamom, coriander.

PRODUCTION Born from the brilliant mind of Craig Allison and head distiller Peter Smith, both, as you will no doubt notice in this elegant London Dry Gin, are lovers of Tonka beans. All the botanicals are poured into a neutral grain spirit about 12 hours before distillation in direct infusion. Distillation always starts at 7am and lasts for 12 hours, with a yield of about 60 liters of 82% abv gin.

NOTES Allison, who had always been a great gin enthusiast, ran a Gin Palace in the heart of Norwich, until the opportunity arose to purchase a 120-liter pot still, which he placed in the nearby pub, The Ten Bells. The distillery has since moved to just over a mile away, increasing its production capacity by 4 times.

Nominated among the Best London Dry at the 2017 World Drinks Awards.

Bullards Gin was launched in 2016.

Founded in 1837, Bullards became a famous brewery, serving the city for 130 years. The anchor featured on the label is taken from the brewery's logo.

RELEASE Bullard's Strawberry & Black Pepper Gin: 40% abv. With strawberries, black pepper, banana.

Bullards Old Tom Gin: 42.5% abv.

G&T GARNISH Lemon or orange zest, thin slice of apricot, strawberry wedge.

COCKTAILS British Spicy Fifty, The Twentieth Century Cocktail, Bee's Knees.

BY THE DUTCH GIN
42.5% ABV

WEB www.bythedutch.com

STYLE Distilled Compound Gin

ORIGIN Schiedam – Holland

BOTANICALS 8 stated. Juniper, nutmeg, coriander, blood orange, lemon peel, cinnamon, cardamom, bay leaf.

PRODUCTION The project was born in 2015 with a genever, while the gin came out just two years later, distilled by the Zonneveld Beverage Company at the House of Herman Jansen Distillery in Schiedam. Each individual botanical is steeped in neutral corn alcohol for variable infusion times (up to 15 days) based on the botanical type, and then distilled separately in a small batch pot still.

NOTES The rope used for packaging is some 2.5 meters long.

The label resembles a vintage newspaper, featuring a considerable amount of product information.

This gin possesses a small "secret" ingredient, a 2% malt wine, used to produce a triple-distilled genever in a copper still.

RELEASE By The Dutch Genever: 38% abv. Made from a 1942 recipe, in perfect old style, in which the percentage of malt wine (consisting of equal parts corn, rye and barley malt) is very high. A distillate of this is added, consisting of juniper berries and botanicals like hops, cloves, star anise, licorice, ginger, citrus fruits and others not stated.

The first distillation of barley, rye and corn malt happens in a Ruwnat pot still; the second distillation in a Enkelnat pot still; the third is column distillation with juniper; redistillation of a portion of the malt wine with botanicals; redistillation of part of the malt wine without botanicals; redistillation of the final part of malt wine with juniper berries; lastly, the blending.

G&T GARNISH Citrus peel or fresh bay leaves.

COCKTAILS Martini Cocktail 1900 Harry Johnson's, Paradise, Abbey Cocktail.

CANAÏMA GIN
47% ABV

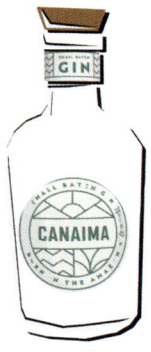

WEB www.canaimagin.com

STYLE Distilled Compound Gin

ORIGIN Lara State – Venezuela

BOTANICALS 19. Ten are from the Amazon (all hand-picked in collaboration with indigenous people of local communities who live freely on the territory), one from Lara (where the distillery is located) and eight traditional botanicals. From the Amazon are: cashews, moriche palm fruit, copoazù, seje, tùpiro, acai, ponsigué, pijiguao, cocuro, maracuja. From Lara: semeruco (an area exotic fruit also known as "acerola"). Among the traditional: juniper, sweet and bitter orange peel, angelica root, cumin, red grapefruit peel and white grapefruit peel.

PRODUCTION Presented in 2019, this product, which harks back to a new Gin Concept, is made in 500-liter batches. The various botanicals are treated with different techniques: from maceration to infusion in neutral grain alcohol for a period ranging from 2 to 7 days, depending on the different concentration of the aroma. They are then distilled separately in traditional copper pot still small batches, to then be carefully combined into a single blend.

NOTES This Super Premium Gin was born in the Amazon, a land where the exceptional bioclimatic conditions have provided access to unexplored botanical varieties in the world of gin.

The name derives from the very famous Canaïma National Park: a protected area located in Bolìvar, Venezuela, and declared a Unesco World Heritage Site in 1994.

For the cultivation and harvesting of local botanicals, a group of indigenous peoples, one of Venezuela's oldest, was called upon: the Pemon, still a very numerous group and one connected to local divinities, divinities who, according to their beliefs, would hide in the grassy lands at the top of the Tepui Mountains (the sacred mountains, forbidden to mortals, as they are inhabited by the Mawari, spirits of the ancestors).

The project is part of The Tierra Foundation, a sustainability promotion strategy for the indigenous tribes of Venezuela.

To combat deforestation in the Amazon, producers collaborate with the Business Forests program of Saving the Amazon, which supports indigenous communities by creating green businesses, which plant, seed, and look after local trees (then photographed and georeferenced). Their images are visible on the website www.savingtheamazon.org.

G&T GARNISH Amalfi lemon peel, pink grapefruit peel, small bunch of aromatic white grapes, passion fruit wedge.

COCKTAILS Gin Fizz, Gin Julep, Bramble.

CARDINAL GIN
42% ABV

WEB www.southernartisanspirits.com

STYLE London Dry Gin

ORIGIN Kings Mountain, North Carolina – U.S.

BOTANICALS 11 stated. Juniper, coriander, angelica, cardamom, grains of paradise, iris, cloves, frankincense, apricot seeds, spearmint, fennel seeds.

PRODUCTION First produced in 2012 by Southern Artisian Spirits in the foothills of the Blue Ridge Mountains, using exclusively organic botanicals. Unique in its kind, as it also includes frankincense among its components, a botanical almost never considered by other producers. Small batch distillation technique.

NOTES Entirely organically produced.

RELEASE Cardinal Gin Barrel Rested: 42% abv.

G&T GARNISH Sorrento lemon peel, fresh mint bunch.

COCKTAILS Southside, American John Collins, Gin Julep.

CITADELLE GIN
44% ABV

WEB www.citadellegin.com

STYLE Distilled Gin

ORIGIN Château de Bonbonnet, Cognac – France

BOTANICALS 19. Juniper, fennel, violet, cassia, cardamom, orange and lemon peel, iris, cinnamon, star anise, coriander, angelica, grains of paradise, almond, nutmeg, Java pepper, cumin, iris, winter savory.

PRODUCTION Owned by the giant Maison Ferrand, which moved production from Dunkerque to Cognac, this gin starts with a triple distillation of French wheat. Subsequently, all spices are infused in alcohol for a maximum of 3 days, then distilled over a high flame for 12 hours in a 2,000-liter copper cognac still—rather rare in the world of gin.

NOTES The original recipe was developed by a Dunkirk distillery in 1771. In 1775, Louis XVI authorized two Frenchmen to open a distillery to produce *genièvre* at the Citadelle. The two founders had 12 copper stills made for the distillery, presumably the first of their kind in France. At the end of the 1980s, Alexandre Gabriel decided the time had come to distill gin and, after discovering the history of Citadelle, he decided that the name and a distillation similar to the original would represent something novel and diverse in the world of gin.

The official production, a first batch that included some of the botanicals noted in 1775, began in 1995, after 6 years of Gabriel's efforts to address some French bureaucratic challenges while trying to obtain the necessary permits. Before him, in fact, nobody had ever used Charentais stills for making gin in the cognac area. However, his tenacity paid off and he succeeded, thanks to some historical documents found that cite the use of such stills for gin production during the stoppage period of cognac production (from April to September).

RELEASE Citadelle Reserve Gin: 44% abv. Aged for 6 months in 12-year-old oak barrels, it was born in 2008 under the supervision of *maître de chai* Frederic Gilbert. Aging takes place in 3 different barrels, ex-Cognac, ex-Pineau de Charentes and American. The result is a blend of these 3 different wood barrels produced in a "bottomless" Solera vat.

G&T GARNISH Wild fennel flowers, Garda lemon peel, star anise.

COCKTAILS Last Word, French 75, John Collins.

COPENHAGEN DRY GIN
44% ABV

WEB www.copenhagendistillery.com

STYLE London Dry Gin

ORIGIN Copenhagen – Denmark

BOTANICALS 1. Juniper.

PRODUCTION The first thing worth noting is that the base distillate is not from grain but rather from a mead, and only a single botanical, juniper, to which a propolis distillate is added, obtained by dissolving propolis in alcohol before distilling it.

NOTES Henrik Brinks is the owner and master distiller of this newly established micro-distillery, located within a wonderful rustic building dating to 1749 (originally a ceramics factory) near Copenhagen. Here, one of the first breweries in the capital city was born. Henrik not only restored the building, abandoned for many years; he also personally designed the 25-liter stainless steel alembic.

RELEASE Copenhagen Aquavit Dill: 40% abv.

Copenhagen Long Pepper Snaps: 41% abv.

Copenhagen Orange Gin: 44% abv. Sweet and bitter oranges, chilli, sweet plum, cardamom, juniper.

Copenhagen Bay Leaf Gin: 45% abv. This gin was created to commemorate their certification as an organic producer. Six botanicals not stated, a base of gin with juniper, laurel, allspice, lemon, angelica, and apparently cranberries. The alcohol base is obtained from organic wheat produced in the still named "Little Bertha".

Copenhagen Oak Gin: 42% abv. A base of juniper, orange peel and pepper. Left to rest in small barrels that previously contained sherry. Selected as Best Cask Gin in 2017.

G&T GARNISH Sorrento lemon peel, Pink Lady apple slice.

COCKTAILS Bennett Cocktail, Bee's Knees, Pink Gin.

COPPERHEAD GIN
40% ABV

WEB www.copperhead.be

STYLE London Dry Gin

ORIGIN Belgium

BOTANICALS 5. Juniper, angelica, cardamom, orange peel, coriander seeds.

PRODUCTION Created in May 2014 by Yvan Vindevogel, a former pharmacist, and distilled at Filliers Distillery in small batches using the traditional London Dry method.

NOTES Mr. Copperhead was an alchemist obsessed with searching for the elixir of life.

The two snakes drawn at the base of the bottle recall the symbol used to indicate a pharmacy, along with Asclepius, god of medicine in Greek mythology.

It is possible to buy 3 different aromatic blends that are dosed in drops (maximum 4 for G&T), to vary cocktail character: Aperitivum (based on angelica and grapefruit), Digestivum (with ginger and pink grapefruit) and Energeticum (with guarana and orange).

RELEASE Copperhead Black Batch: 42% abv. London Dry Gin with the particular trait of containing among its botanicals common elderberries (Sambucus Nigra) and Ceylon black tea, popular in South India.

Copperhead Barrel Aged: 40% abv. This version is aged for 3 months in former Pedro Ximénez barrels, and is the result of a collaboration between Vindvogel and Bernard Fillier, owner of the Filliers distillery.

Copperhead Gibson Edition: 46% abv. The fruit of a collaboration with London's famous gin bar The Gibson, owned by Mariam Beke, a true legend in the world of bartending. The base is a classic Copperhead, with an additional 14 botanicals (especially spices) selected by Beke. These include allspice, bay leaf, cassia, dill seeds, fennel, ginger, mace and pepper, plus an addition of 8-year-old genever, which lends a touch of sweetness without added sugars.

G&T GARNISH Lemon peel, pomelo peel, blackberries.

COCKTAILS Bebbo Cocktail, Negroni, Suffering Bastard.

COTSWOLDS GIN
46% ABV

WEB www.cotswoldsdistillery.com

STYLE London Dry Gin

ORIGIN Cotswolds-Northleach, Cheltenham – United Kingdom

BOTANICALS 9. Juniper, coriander, cardamom, angelica, bay leaf, pepper, lavender, lime, grapefruit.

PRODUCTION A man of finance, New Yorker Daniel Szor moved to London in 2006 with his wife and founded the Cotswolds Distillery. The couple, having fallen in love with the Cotswolds, bought a weekend home in Shipston-on-Stour. In 2012 Szor decided to open the distillery, starting off with whiskey. The stills are branded Forsyths, one of the best makers of whiskey stills, while the gin is produced in an original Holstein still.

Four kilos of spices are used for each distillate, with a resulting highly intense flavor. The botanicals are macerated for several hours in neutral grain alcohol, then distilled for about 7 hours in a 500-liter still that is filled to only two-thirds capacity. Production is just 400 bottles.

NOTES Only locally sourced botanicals gathered in the region are used.

At the start of his adventure, Szor asked for support from the nearby Bruichladdich Distillery, in Scotland. Their famous master distiller, Jim McEwan, then put him in contact with veteran master distiller Harry Cockburn, with whom Szor began to construct his immaculate distillery, with careful attention to every detail.

G&T GARNISH Lavender, lime zest or yellow grapefruit peel.

COCKTAILS Alaska, Martini Cocktail, Gimlet.

CUBICAL PREMIUM 40% ABV
AND SUPER PREMIUM 45% ABV

WEB www.gincubical.com

STYLE London Dry Gin

ORIGIN Cadiz, Jerez de la Frontera – Spain

BOTANICALS 15 stated. Juniper, coriander seeds, thyme, angelica root, orange and lemon peel, orris root, cinnamon, peppermint, chamomile, licorice root, cassia, almond, anise, sweet orange, Buddha's hand.

PRODUCTION Distilled by Langley Distillery, founded in Birmingham in 1920, and owned by Williams & Humbert. Premium undergoes 3 distillation processes, while the Super Premium has 4.

NOTES W&H was founded in 1877 by Sir Alexander Williams, expert and enthusiast of Jerez products, and Arthur Humbert, an international relations specialist. Recognized as one of the most prestigious wineries in Jerez, and world-renowned for the sherry production.

Buddha's Hand, scientifically known as Citrus Medica of the Sarcodactylis variety, is a fruit that grows in segments, each segment developing separately from the others, and each creates its own particular rind, hence the various "fingers" of Buddha's hand. The fruit has a thick skin, a small amount of pulp, and is sometimes seedless, with a sweet-tasting white inner pith.

RELEASE Gin Cubical Kiss: 37.5% abv. Triple distilled. Same botanicals used for the Premium, with added berries such as strawberries, raspberries and blueberries. Scores 94 points on the Gold Taste Wine Up Club TOP 100 points.

Gin Cubical Mango: 37.5% abv. Triple distilled. Same botanicals as Premium with the addition of sweet orange, bergamot and mango.

G&T GARNISH Citron peel, fresh mint sprig, fresh chamomile flowers.

COCKTAILS Florodora, Seventh Heaven, Gin Crusta.

DEATH'S DOOR GIN
47% ABV

WEB www.deathsdoorspirit.com

STYLE Wisconsin Dry Gin

ORIGIN Middleton, Wisconsin, Washington Island – U.S.

BOTANICALS 3 stated. Wild juniper harvested on Washington Island, fennel seeds, coriander.

PRODUCTION The juniper used is organic and grows wild on the island, while the coriander and fennel are harvested in the state. A blend of grains is used, namely wheat and barley. The yeast for fermentation is a Champagne yeast. Distillation takes place in column stills in 3 steps, and the same process is followed for the corn. Their vodka is the result of the union of these two spirits. The botanicals pass through a botanical extractor, then added before the end of the third distillation in a pot still. In perfect London Dry Gin style.

NOTES Washington Island has an area of 60 km^2 and boasts about 1,100 km of coastline.

In 2005, brothers Tom and Ken Koyen, supervised by the Michael Fields Institute, planted two hectares of wheat, guided by an unwavering belief in the abandoned area. Death's Door Distillery, founded by Brian Ellison based in Middleton, Eastern United States, opened its doors on June 4, 2012 and is now the largest in Wisconsin.

The name "Death's Door" describes the area located between Door Country and Washington Island, which was originally called "The Waterway". For the French, however, it was known as the *Port de Morts* (harbor of death), likely assigned to deter other traders from coming into the area.

The company donates 1% of its revenue to the cleanup of the Great Lakes. In 2015, the company joined Planet, a global cleanup initiative.

Harvesting juniper berries is an extremely long, exhausting practice, so Death's Door opens its doors during the first week of November (peak harvest time on Washington Island) and invites all friends and fans to participate in a collective juniper harvest.

RELEASE Death's Door Vodka: 40% abv. American Small Batch Vodka.

G&T GARNISH Garda lemon peel, fresh mint springs.

COCKTAILS Gin Mule, Corpse Reviver No. 2, American Gin Fizz.

DODD'S GIN
49.9% ABV

WEB www.londondistillery.com

STYLE Distilled Compound Gin

ORIGIN London – United Kingdom

BOTANICALS 8. Juniper, bay leaf, black cardamom, green cardamom, raspberry leaves, honey from The London Honey Company, angelica root, lime zest.

PRODUCTION Produced in small batches, with entirely organic botanicals that are processed separately. The raw materials used to make the base distillate are also organic and approved by the Soil Association. A significant part is first distilled in a small copper pot still (only 140 liters capacity) named "Christina" (previously called "Christian Carl Still"). The more delicate botanicals are instead distilled "cold" and vacuum-packed in a micro-still named "Little Albion". The two distillations are then combined and left to rest in steel for several weeks before bottling by hand.

The head distiller is Andrew MacLeod.

NOTES Made at the new Battersea Distillery in the heart of London, this gin is in honor of Ralph Dodd, a long-standing entrepreneur who in 1807 unsuccessfully convinced investors to finance a new distillery, one that would produce high quality spirits for the British public. In 2011, more than a century after the last London distillery had closed its doors, Darren Rook and Niky Taylor brought The London Distillery Company to life.

The Greenness of Dodd's labels are printed on carbon neutral paper made exclusively using wind energy. The company has promised to plant two trees for each barrel used, and even the thermal energy used for distillation is recuperated.

The London Distillery Company is located in the old milk warehouse known as "The Cold Room" (formerly a Victorian dairy) in Battersea.

In homage to Ralph Dodd, the label is inspired by trigonometry, geometry and engineering schemes.

G&T GARNISH Garda lemon zest, lime zest, fresh raspberry.

COCKTAILS Clover Club, Florodora, Seventh Heaven.

DRY FLY HANDCRAFTED GIN
40% ABV

WEB www.dryflydistilling.com

STYLE Washington Dry Gin

ORIGIN Spokane, Washington – U.S.

BOTANICALS 7 stated. Juniper from Oregon, Fuji apples, coriander, angelica, lavender, hops, mint.

PRODUCTION Produced at its own distillery, this gin uses local grains from recognized growers, in order to control the entire production process, from field to bottle. Once harvested, grains are immediately taken to a hammer mill of Austrian origin that grinds it to a flour, which is subsequently processed to "accommodate" the yeasts. This process lasts about 5 hours, while the room-temperature fermentation lasts one week. The distillations (from 2 to 4) are carried out in a 450-liter Christian Carl copper still. The gin comes from a neutral grain distillate infused with botanicals and then redistilled. On average, it takes about 10 days from grain to bottle.

NOTES A small distillery in Washington State, Dry Fly produces a full range of spirits using locally sourced wheat and plants. The distillery was built by Dan Poffenroth and Kent Fleishmann, great friends and fly fishing enthusiasts (which explains the brand logo).

RELEASE Dry Fly Barrel Aged Gin: 40% abv. Aged for a year in ex-whisky (wheat) barrels.

Dry Fly Vodka: 40% abv. Made with pure soft winter wheat grown by a family that has been doing this work for some 120 years. Distilled 3 times.

G&T GARNISH Fresh mint, slice of Golden apple, Sorrento lemon peel.

COCKTAILS American John Collins, Aviation, South Side.

EDEN.MILL St. ANDREWS GIN
42% ABV

WEB www.edenmill.com

STYLE London Dry Gin base, Compound for some selections

ORIGIN Fife, Scotland – United Kingdom

BOTANICALS 3 stated. Juniper, coriander seeds, angelica.

PRODUCTION Made in Scotland by St. Andrews Distillery & Brewery, and founded by Paul Miller. The neutral corn distillate used as the base for this gin is purchased from outside sources. The distillation system is vapor infusion. The botanicals are suspended in the upper part of the 3 Portuguese still, producing an average of 1,000 liters of gin at 43% abv. Thus, each recipe will be cut based on freshness and availability of the local raw materials used. A single batch takes about 17 hours to make, and involves passing the steam through a slighter amount of botanicals, only 4 kg.

NOTES The name of this gin, whose packaging recalls beer bottles of the past, derives from the nearby estuary. Bottling takes place by hand, about 1,500 per week.

Production takes advantage of seasonality, to develop different editions based on what is offered locally, all in limited edition.

RELEASE Eden Mill Hope Gin: 46% abv. Australian Galaxy hops base, cold inserted after distillation, thus uniting the distillery and the brewery (compound style). Batches are 980 bottles.

Eden Mill Oak Gin: 42% abv. Left to rest in ex-bourbon barrels using the same base as the Hope Gin, except for the hops (compound style).

Eden Mill Original Sea Buckthorn Gin: 42% abv. The name suggests the use of the same-named berry, quite rare in the world of gin. The other characterizing botanicals are lemon balm and citrus peel (London Dry Gin style).

Eden Mill Love Gin: 42% abv. Released in the summer of 2015 with 8 exotic aromas, including rose petals, rhubarb root, elderberries, althaea (marshmallow root), goji berries, raspberry leaves and hibiscus flowers, the last of which is infused after distillation (compound style).

Eden Mill Golf Gin: 42% abv. Launched in the summer of 2016 in conjunction with the Open Golf tournament at the famous St. Andrews course. Primarily characterized by splinters of Carya wood and lemongrass (compound style).

G&T GARNISH Sorrento lemon peel, pink grapefruit peel.

COCKTAILS Clover Club, Singapore Sling, Cardinale.

EDINBURGH GIN
43% ABV

WEB www.edinburghgin.com

STYLE London Dry Gin

ORIGIN Broxburn, West Lothian – United Kingdom

BOTANICALS 11 stated. Juniper, angelica, lemon peel, coriander, iris, lavender, pine buds, lemongrass, mulberry, thistle, heather.

PRODUCTION The distillery produces in 150-liter small batch stills named "Caledonia" and "Flora". Then there is "Jeanie", 1,000 liters, added in 2016 after the opening of the second distillery in Leith, at which point ownership passed to Ian Macleod, a famous producer of scotch whiskey.

NOTES Established in 2010, Edinburgh Gin Distillery was the first gin distillery to open its doors in the Scottish capital.

Two plants are in operation now, one at the end of Princes Street in the port of Leith and the other in The Biscuit Factory in Anderson Place, also in Leith.

The team is led by master distiller David Wilkinson.

Created in homage to the beautiful and wild Scottish coast, Edinburgh Seaside Gin is the result of a collaboration with Heriot-Watt University.

Edinburgh Gin Distillery produces its London Dry Gins from a double distillation, using 3 pot stills. During the first distillation, the heavier and more resinous botanicals, such as juniper, are macerated. The distillate then passes to Flora, in the upper part of which the more delicate botanicals such as flowers, algae and citrus fruits are hung.

RELEASE EG Seaside: 43% abv. Features hand-picked coastal botanicals, scurvy grass, sea oak and ivy.

EG Cannonball: 57.2% abv. Inspired by Edinburgh's naval military tradition, with double the amount of juniper, lemon peel and Sichuan pepper.

EG 1670: 43% abv. The result of an innovative collaboration with the Royal Botanic Garden of Edinburgh, with coriander seeds, licorice root, cardamom and Tasmannia pepperberry (leaves and berries).

EG Apple & Spice: 20% abv. With apple, cinnamon, orange peel, lavender, lemongrass and black mulberry.

G&T GARNISH Fresh lemon peel, lavender, lemongrass piece, red berries.

COCKTAILS Straits Sling, Monkey Gland, Last Word.

ELEPHANT LONDON DRY GIN
45% ABV

WEB www.elephant-gin.com

STYLE London Dry Gin

ORIGIN Hamburg – Germany

BOTANICALS 14. Juniper, mountain pine needles, lavender, sweet orange peel, fresh apple, cassia bark, ginger, allspice berries, devil's claw, dandelion, elderberry, African mugwort, baobab.

PRODUCTION A superior quality gin with an alcohol content of 45% that aims to capture the essence of Africa. The unique blend of 14 herbs and spices includes rare African ingredients and fresh apples that lend the gin a distinctive flavor profile. Maceration for 24 hours, one-shot distillation and production of about 800 bottles per batch.

Elephant Gin is distilled using artisanal methods, with Holstein discontinuous copper stills, in small batches of about 600 bottles.

NOTES Its creators were inspired by their adventures in Africa and the idea of "sundowner", a kind of aperitif that is sipped at sunset, after a day spent in the savannah.

The producers firmly believe in the responsibility of current generations to preserve our planet's natural wonder and beauty. For this reason, Elephant Gin contributes 15% of its earnings to the 3 organizations: Big Life Foundation, Space For Elephants and David Sheldrick Wildlife Trust, involved in the protection of the African elephant, which is at risk of extinction.

The first release came out in 2013.

The distillery was founded by Robin Gerlach, Tessa Wienker and Henry Palmer.

"Every year more than 35,000 elephants die due to ivory poaching; in other words, one every 15 minutes".

Each batch is dedicated to an elephant and takes its name from one (for example "Dionysis").

RELEASE Elephant Navy Strength Gin: 57% abv.

Elephant Sloe Gin: 35% abv.

G&T GARNISH Slice of red apple, thin slice of ginger, lavender sprig.

COCKTAILS Bramble, Gin Fix, Monkey Gland.

FERDINAND'S SAAR DRY GIN
44% ABV

WEB www.ferdinandsgin.com

STYLE London Dry Gin

ORIGIN Wincheringen – Germany

BOTANICALS 30 stated, including juniper, lemon thyme, blackthorn, rose hips, angelica, hop blossom, rose petals, almond shell, coriander, ginger, lavender and rosehip.

PRODUCTION Produced by the Avadis Distillery. The distillation process begins by creating the base spirit from harvested grain, which reaches an alcohol content of 90-95% abv after several distillations, cut with water before adding the 30 carefully selected organic botanicals harvested from the small valley nearby Konzer Tälchen. The lighter aroma botanicals are also added inside the steam infusion chamber of the still, to avoid dispersing the essences. The two methods (maceration and infusion) are used during a single distillation, which is very rare. This is followed by the addition of a certain amount of Schiefer Riesling, then a rest of 4 weeks, and another addition of water to bring the alcohol content to 44%.

NOTES Ferdinand's Saar Dry Gin owes its name to the Royal Prussian District Forester Ferdinand Geltz, one of Germany's best vineyards, located on the large site of the Saarburger Rausch, widely known for its rich mineral soil.

Producers are Andreas Vallendar (master distiller) and his Avadis distillery in Wincheringen, Dorothee Zilliken from the VDP Forstmeister Geltz-Zilliken estate in Saarburg, and the marketing company Capulet & Montague in Saarbrücken.

The Vallendar family has been distilling spirits for several generations on their estate, which was founded in 1824 in Wincheringen.

The shape of the bottle resembles wine bottles and is sealed with a cork and beeswax.

RELEASE Ferdinand's Saar Quince Gin: 30% abv. Cut with Rausch Kabinett 2011, a sloe gin style in which, however, sloe is replaced with fresh quince.

Ferdinand's Saar Quince Gin Goldcap: 49% abv. Limited edition (one production per year) with a base of acacia, cocoa beans, dehydrated Riesling grapes, juniper, Mirabelle plums, pears and the addition of a special wine, Auslese 2010.

Ferdinand's Saar Cask Strength Gin: 66.6% abv.

G&T GARNISH Lemon peel, thyme sprig, small grape bunch (preferably Riesling).

COCKTAILS French 75, Corpse Reviver No. 2, Gin Julep.

FILLIERS 28 GIN
46% ABV

WEB www.filliersdrygin28.be

STYLE Belgian Gin, Distilled Compound Gin

ORIGIN Bachte-Maria-Leerne – Belgium

BOTANICALS 28 stated, including juniper from Italy, coriander from Bulgaria, lavender blossom from France, orange blossom from Spain, Belgian white hop blossom, cardamom, angelica root, gentian root, ginger, calamus, lemon peel, lime and orange, aloe vera, chicory root, cinnamon, coriander, allspice, dandelion flowers and raspberries.

PRODUCTION The production (distilled approximately twice a week) takes place entirely through a 200-liter Porty old still and two other 500-liter pot stills. Many of the herbs are distilled separately and then blended for the final product. For example, oranges are left to macerate in alcohol for 3 days separately, then distilled.

NOTES Number 28 is a reference to botanicals, but also to the year Firmin Filliers created the distillery's first gin recipe. Already considered a "distillery of essence", it is also noted for the bottle shape, which emulates a pharmaceutical bottle.

It wasn't until 1880 that Filliers had the chance to install a new, modern steam engine distillation plant. Doing so allowed them to become competitive on the European market.

Today Filliers is one of the oldest distilleries in Europe, along with Dutch Bols, and is also among the largest existing producers of genever, in terms of both its portfolio of products and for its status as a supplier to other companies. From a combination of rye and malted barley, together with corn and wheat, a moutwijn is made, something that makes this company unique in its kind—as a direct producer of both base alcohols and their own final preparations.

RELEASE Filliers 28 Pine Blossom Gin: 42.6% abv. With added Scots pine from the Arctic, which is macerated for 3 weeks.

Filliers 23 Barrel Aged: 43.7% abv. Aged for 4 months in 250-liter French oak barrels.

Filliers Dry Gin 28 "Sloe Edition": 26% abv. Uses only wild Allegheny blackthorn, in the amazing amount of 1 kg for every 2 liters of gin.

Filliers Dry Gin 28 "Tangerine Edition": 43.7% abv. Made with premium Valencia mandarins.

G&T GARNISH Lemon or orange peel, red berries, lavender sprig.

COCKTAILS Cardinale, Ramos Gin Fizz, Clover Club.

FOREST DRY GIN
40% ABV

WEB www.forest-spirits.com

STYLE Distilled Compound Gin

ORIGIN Antwerp – Belgium

BOTANICALS 7 + release. Juniper.

PRODUCTION The base used for all the processes is a grain distillate obtained from a quadruple distillation in a copper alembic pot still.

The botanicals and fruit are first macerated, then distilled separately, then combined in the base distillate.

NOTES Company is owned by Jurgen Ljcops, expert sommelier and owner of The Glorius, a Michelin-starred restaurant near Antwerp.

Their philosophy is to obtain pure juniper-based distillates, to combine according to the seasons using ingredients produced "at home".

The 1.5-hectare estate, in fact, grows its own bergamot, coriander, angelica, pears, apples and lavender.

RELEASE Forest Autumn Gin: 42% abv. With pear, lavender and mandarin.

Forest Spring Gin: 42% abv. Floral with roses, lychee and wild strawberries.

Forest Summer Gin: 45% abv. Blood orange, bergamot flower and ginger.

Forest Winter GIN: 45% abv. Apple, coriander and citrus.

Forest Quercus Gin: 42% abv. The Autumn version is aged for one year in ex-Sauternes oak barrels hand-painted by the artist Valère Maenhout.

Forest All Season Gin: 38% abv. Produced with over 24 different botanicals.

Forest Valentine Gin: 45% abv. Produced from a base of 15 aphrodisiac herbs grown on the farm. 100% organic.

Forest Dry Gin Anno Domini XV: 57% abv. The only product in perfect London Dry Gin style. Only 480 Navy Strength style bottles: 50 grams of juniper per liter and a pinch of Belgian hops.

G&T GARNISH Pink grapefruit peel, thin slice of ginger, edible rose petals.

COCKTAILS Gin Smash, Gin Collins, Gin Old Fashioned Seasonal.

FOUR PILLARS RARE DRY GIN
41% ABV

WEB www.fourpillarsgin.com

STYLE Australian Dry Gin

ORIGIN Yarra Valley – Australia

BOTANICALS 10. Juniper, coriander, cardamom, Australian myrtle, Tasmanian pepper leaves, cinnamon, lavender, angelica, star anise, fresh orange pulp.

PRODUCTION The still used is a Carl handcrafted still (the oldest producer of stills in Germany, with a limited production of not more than 25 per year) made in Stuttgart. It has a capacity of 450 liters and a basket to contain the orange pulp. Each batch requires 7 hours of distillation to produce approximately 150 liters of high-proof gin, for approximately 460 bottles per batch. The wheat used for grain spirit comes from the south coast of New South Wales. Subsequently, the alcohol content is lowered to 30% abv and 450 liters are then poured into the pot still Wilma's Belly, where the 9 botanicals will be inserted. The oranges are instead placed in a special basket located in the upper part of the still.

NOTES Born from the minds of Stuart Gregor and Cameron Mackenzie along with guru Matt Jones, Four Pillars was launched in 2013 as one of Australia's first artisanal gins.

The distillery is based in the Yarra Valley near Melbourne and is a project partially funded by Pozible, a crowdfunding platform. They exceeded their initial target of AU$10,000 on the second day of the campaign.

In reality, the initial project was aimed at tonic water. Only later did they decide to shift their resources and attention towards the production of a quality gin that reflects the territory.

RELEASE Four Pillars Bloody Shiraz Gin: 37% abv. Born in 2015 from the union of Rare Dry Gin with some of the best Shiraz grapes in Australia.

Four Pillars Navy Strength Gin: 58.8% abv.

Four Pillars Spiced Negroni Gin: 43.8% abv. Extra quantities of cinnamon and Tasmanian pepper leaves are added to the classic Rare Dry Gin. Grains of paradise (an exotic spice from West Africa), organic blood oranges and ginger have also been added.

Four Pillars Barrel Aged Gin: 43.8% abv. Aged in barriques that previously contained chardonnay. Subsequently, the Sherry Cask Gin and the Australian Christmas Gin, aged in muscat barrels, were also released.

G&T GARNISH Amalfi lemon peel, fresh orange peel and slices, sprig of lavender.

COCKTAILS Gin Cobbler, Florodora, Corpse Reviver No. 2.

FRED JERBIS GIN43
43% ABV

WEB www.fredjerbis.com

STYLE Italian Distilled Gin

ORIGIN Polcenigo (Pordenone) – Italy

BOTANICALS 43 botanicals, all of Italian origin, including juniper, angelica, lemon, bitter orange, mandarin, thyme, lavender, mint, anise, fennel, mountain pine, lemon balm, iris, master wort, winter savory, clary sage, yarrow, saffron, orange flowers, bitter almond, elderberry, dandelion, wormwood, thistle, marjoram, lemon and mandarin peel, dill, caraway, mint, gentian, bay leaf, common centaury, coriander, licorice, wild thyme and hyssop.

PRODUCTION Launched in 2014.

Producing GIN43 requires 5 extraction methods: alcoholic distillation, steam distillation, cold maceration of the dried plants, cold maceration of the fresh plants, and hot maceration (digestion).

All this takes place in Polcenigo, inside the historic former pharmacy, now a laboratory. Each botanical is processed individually according to the best specific extraction, and only then is the recipe created, using a high percentage of juniper. The straw color of the product also underlines this, accentuated by the cold infusion of orange blossom and saffron.

NOTES The territory is a fundamental aspect for this product. Almost all the botanicals are grown locally. The strength of this distillery lies in their network of established collaborators, with micro-productions consistently guaranteeing the highest quality ingredients.

To make this gin, the producers were inspired by some texts on Italian liqueur tradition.

Neither of the company owners came from distilling or entrepreneurial backgrounds. Federico has long been a recognized and established barman, while Massimo came from a background in marketing.

RELEASE GIN7 Single Barrel: 43% abv. Created with only 7 botanicals (juniper, angelica, coriander, orange, mandarin, lemon balm and iris) and aged in new acacia barriques produced by the skilled hands of a Friulian cooper.

G&T GARNISH Lemon peel, fresh mint sprig, fresh thyme sprig.

COCKTAILS Old Tom Gin Cocktail, Gin Julep, Tom Collins.

FYNODEREE MANX DRY GIN
43% ABV

WEB www.fynoderee.com

STYLE London Dry Gin

ORIGIN Isle of Man – United Kingdom

BOTANICALS 15 stated, including wild elderberry, sloes, blackberries, fresh ginger, cloves and cinnamon.

PRODUCTION Produced by The Fynoderee Distiller and a project born from 3 founders: Tiffany and Paul Kerruish and master distiller Gerard Macluskey.

A mixture of neutral grain spirit and water is poured into a 300-liter copper pot still, in which the botanicals (not previously macerated) are added. Distillation takes place with a very "old school" open gas fire. The master distiller has previously worked for Masons Distillery and the iconic Tanqueray Gordon & Co. Each batch yields approximately 210 bottles.

NOTES Located is in the northern part of the island where juniper is being reintroduced.

It took Macluskey just 9 months to finalize their first release.

The first edition produced was the Winter Edition in November 2017.

The husband and wife team share a surname with the Kerruish clan, a well-known name on the island as it is linked to a fairy tale in which the beautiful Kitty Kerruish made Fynoderee, a mythical creature (a kind of satyr), fall in love with her.

The distillery promotes the reuse of glass: customers show up with empty bottles and fill them.

RELEASE Fynoderee Spring Edition 2018: 43% abv. Distilled with fresh coriander, juniper, mint, lemon, verbena, coconut and ginestra (broom) flowers.

Fynoderee Winter Edition: 43% abv. Distilled with winter berries, including hand-picked elderberries and blackberries, then sloes and cinnamon.

Fynoderee Summer Edition: 43% abv. Distilled with fresh local strawberries, citrus fruits and summer spices from the island.

Fynoderee Autumn Edition: 43% abv. Distilled with local apples, rowan berries, rose hips and plums.

Fynoderee Kerala Chai Edition: 43% abv. In collaboration with the third generation of a spice merchant family and chef Kumar Menon of Leela's Kitchen on the Isle of Man, this gin is distilled with hand-picked Himalayan juniper and Assam tea leaves, combined with Chai Masala spices imported from chef Menon's birthplace: the Kerala region of India.

G&T GARNISH Orange peel, lemon peel, red berries.

COCKTAILS Bramble, Gin Cobbler, Gimlet.

G' VINE GIN (FLORAISON AND NOUAISON)
40% AND 43.9% ABV

WEB www.g-vine.com

STYLE Distilled Gin

ORIGIN Merpins, Charente – France

BOTANICALS 10. Macedonian juniper, Indonesian cassia and nutmeg, Nigerian ginger, Chinese licorice, green cardamom from Guatemala, coriander from Bulgaria, cubeb berries from India, and lime from Colombia. Also featured is inflorescence, also known as a vine blossom (or fleurs de vigne). Hence, flowering and flora are reflected in the name "floraison", while "nouaison" refers to the first pod used, the part immediately following the flower (called the "setting" in English).

PRODUCTION Produced directly from neutral alcohol obtained from Ugni Blanc, a typical grape that comes from the famous cognac-producing region.

Production begins, just like for wine, in September, when the grapes are harvested, turned into wine and distilled (column), with a result of 96.4% abv.

Once harvested, the flowers are immediately left to macerate in the grape alcohol for a few days. All botanicals are macerated and distilled separately in order to extract the best qualities from each. The various distillates are then combined with the alcohol and flower maceration, adding a portion of neutral alcohol again. Finally, it is redistilled all together in a copper pot still called "Lily Fleur".

NOTES Gathering the inflorescences takes place in June. As these blooms are only available for 10 to 15 days, if they are not gathered at this time then one must wait until the following year. The harvest is carried out entirely by hand. The master distiller is Jean-Sébastien Robicquet, while the owner company is EWG (EuroWineGate) Spirits & Wines.

G&T GARNISH Garda lemon peel, lime zest, small bunch of aromatic, in-season white grapes, thin slice of ginger.

COCKTAILS Gin Cobbler, Gimlet, French Martini.

GASTRO GIN&JONNIE
45% ABV

WEB www.gastrogin.com

STYLE Apparently a distilled compound

ORIGIN Schiedam – The Netherlands

BOTANICALS 16. Juniper, lemon peel, grapefruit and orange, angelica, licorice, lemon verbena, fennel flowers and seeds, cumin, cardamom, long pepper, allspice, Sichuan pepper, Sarawak pepper, Voatsiperifery pepper.

PRODUCTION Distilled at the Onder de Boompjes, the second oldest distillery in The Netherlands and born in 1658, when the Steffelear family started producing genever and korenwijn.

NOTES Created in collaboration with 3 Michelin star chef Jonnie Boer of the De Librije restaurant in Zwolle. Boer arrived at the restaurant as head chef in 1989, and in 1992 purchased it together with his wife and business partner, Thérèse Boer. A year later he won his first Michelin star, then the second in 1998, and, in 2004, he achieved the highest recognition among chefs, a third Michelin star.

Great attention is paid to the absolute freshness and quality of every single botanical used. All citrus fruits are peeled exclusively by hand.

G&T GARNISH Yellow grapefruit peel, Garda lemon peel, fennel flower, verbena leaf.

COCKTAILS Dirty Martini, Red Snapper, Negroni.

GERANIUM GIN
44% ABV

WEB www.geraniumgin.com

STYLE London Dry Gin

ORIGIN Denmark, but also produced in Birmingham, United Kingdom

BOTANICALS 9 stated + 1 secret. Savin juniper, coriander, lemon and orange peel, angelica, iris, cassia, licorice, essential oils of Danish grains.

PRODUCTION Owned by Hammer & Son LTD but distilled at Langley Distillery in Birmingham, it launched in September 2009 in Denmark as a new gin concept. A creation of Henrick Hammer, member and recognized judge of the IWSC (International Wine & Spirit Competition, founded in London in 1969), who has always been fascinated by the use of geranium on a therapeutic level. He believes he can increase the spectrum of botanicals present in the world of gin.

Botanicals are infused for 48 hours in neutral grain alcohol. Distillation is in a copper pot still called "Constance".

NOTES Henrik conducted a chemical analysis of geranium, discovering that its significant oils (geraniol, geranyl formate, linalool, rose oxide, citronellol) are commonly found in fruits, vegetables and spices. They are therefore already present, for example, in juniper and lemon.

Geraniol has numerous properties: it stimulates the adrenal cortex, helps balance the nervous system, relieves depression and anxiety, and has regulatory effects on the hormonal system.

Henrik's father, Hammer Hudi Senior, was a chemist-pharmacist specializing in the production of essences for perfumes and medicines, and who worked in the Danish industry for years. With his son, he bought a 5-liter copper alembic and began experimenting with how to best bind geranium. Their specific idea was to craft an authentic London Dry Gin based on geraniol. For this, they turned to one of the best English distilleries around.

A significant problem arose related to geranium. Until then, in fact, aromatic oils were extracted with water vapor and not in alcohol. However, thanks to Hammer's knowledge, they managed to find the solution, using a small 5-liter still.

The geranium is grown and harvested by Henrik in Denmark, and is then shipped to Langley Distillery, where it is used fresh.

RELEASE Geranium Gin 55°: 55% abv. Born to celebrate the 5th anniversary of the birth of this production. What's more, 55 is also the exact latitudinal parallel that crosses the land of its origin.

G&T GARNISH Geranium shoots, Ligurian lemon zest, grapefruit peel.

COCKTAILS Southside, Florodora, Pegu Club Cocktail.

GIASS DRY GIN
42% ABV

WEB www.giassgin.com

STYLE London Dry Gin

ORIGIN Milan – Italy

BOTANICALS 18, subdivided into 6 notes/categories:

Classic base: juniper berries, coriander seeds, angelica root.

Fruity notes: Golden apple and orange peel, dried naturally at low temps.

Floral notes: rose petals, chamomile flowers, violet flowers, orange flowers, karkadè.

Mineral notes: cypress or earth almond (tiger nut), mint leaves, fennel seeds.

Citric notes: citrus verbena leaves, cardamom seeds, lemon balm leaves.

Woody notes: thyme, cassia.

PRODUCTION Giass's authentic recipe was entirely conceived by the 5 founding members. The first tests took place in 2015. The infusion was made up of vodka, to which botanicals were added, and the rudimentary "distillation" instrument was a dishwasher, for one hour at about 70° C. Duration and high temperatures allowed for lowering infusion times and increasing the quality of the final result.

Next, a 3-liter still was purchased on Amazon, and with this the idea of creating the first London Dry Gin in Milan was born.

After 6 months of testing, Giass Gin was launched: a contemporary recipe with 18 botanicals, distilled according to the traditional London Dry method.

Today Giass Gin is made in a discontinuous still. The procedure involves a combined maceration (which takes place in Trentino) of all the botanicals for 72 hours.

NOTES Giass, meaning "ice" in the Milanese dialect, is the first and only London Dry Gin in Milan.

Giass is bottled in an elegant silk-screened bottle, called the "Vecchia Farmacia", or "Old Pharmacy".

The Giass logo recalls the green dragon of the *vedovelle*, characteristic Milanese public drinking fountains.

RELEASE Navy Strength: 57% abv. Same recipe as their London Dry, only with increased botanical proportions.

G&T GARNISH Lemon or yellow grapefruit peel, slice of Granny Smith apple, fresh mint sprig, fresh thyme.

COCKTAILS Aviation, French 75, Floral Gin Fizz.

GIN DEL PROFESSORE

WEB www.delprofessore.it

STYLE Cold Compound Gin

ORIGIN Piedmont – Italy

BOTANICALS 18 stated. Juniper, angelica, chamomile, lavender, orange, zedoaria (white turmeric), rose, lemon, orange, tansy, cinnamon, cassia, vanilla, cocoa beans, cloves, coriander, elderberry, allspice.

PRODUCTION Produced at the Quaglia Distillery in Castelnuovo Don Bosco (Asti) in Piedmont. The production system perfectly recalls a bathtub gin style, reinterpreted thanks to the high quality of the area surrounding the distillery, an area famous around the world for its variety of herbs and spices.

The production starts with a neutral base distillate from grains, to which wild Umbrian juniper is added, and is then redistilled. It is a kind of London Dry Gin in all respects, subsequently used as a base for making cold infusions, which differ in terms of both botanical and alcoholic volume. It is then carefully filtered to preserve each precious element. Given the high level of craftsmanship involved, 3 months pass before bottling. Everything is processed in single batches and each production reveals differences in freshness, reflecting the raw material used.

NOTES Monsieur and Madama contain the same types of botanicals, yet they are handled in different quantities and infusions, while Crocodile has a different botanical portfolio (excluding juniper), consisting of 10 spices, including citrus fruits, coriander, elderberry and allspice.

Behind this wonderful portfolio of products, including the property itself but also, and primarily, their research, is the impressive work of some of the finest producers in this field: Leo, Roby, Tony and Ale, who are the founders of the Jerry Thomas Speakeasy (www.thejerrythomasproject.it) and Punta (www.lapuntaexpendiodeagave.com). Both are well-known Roman locales of international calibre.

RELEASE Monsieur: 43.7% abv.

Madama: 42.9% abv.

Crocodile: 45% abv.

G&T GARNISH Fresh lemon or orange peel, dusting of fresh chamomile, dusting of Tonka beans.

COCKTAILS The Pendennis Cocktail, Martinez, Alaska.

GRANIT BAVARIAN GIN
42% ABV

WEB www.granitgin.com

STYLE Bavarian Dry Gin

ORIGIN Hauzenberg, Bavaria – Germany

BOTANICALS 28 from organic farming: some come from the Bavarian forest (gentian, lemon balm, wild liquorice and alpine fennel), which lends Granit its original character, while others are classic (juniper, lemon, ginger, cardamom, coriander and pepper).

PRODUCTION Production began on Valentine's Day, 2014.

The dry botanicals are macerated for 12 hours, then the fresh botanicals are added and everything is immediately distilled. Distillation takes place slowly, over about 12 hours, in a small, wood-heated 150-liter still. Finally, the gin is stored for a few months in traditional terracotta containers, before being filtered with granite (from which it takes its name). For the production of Granit, an old device called an "Oxy-Esterator" (created in the 1960s, housed in the Penninger Museum) was put back into use. It serves to filter the gin using granite of various sizes, removing most of the heavy oily substances without compromising the distillate.

NOTES Each bottle is sold with a piece of granite included, which can be placed in the freezer and used as an ice cube (to avoid dilution)—a detail that, like the product's name and label, recalls eastern Bavaria's long granite mining tradition.

G&T GARNISH Fennel flowers, Garda lemon peel, ginger slice.

COCKTAILS Gin Smash, Pegu Club Cocktail, Aviation.

GREEN HAT GIN
41.1% ABV

WEB www.greenhatgin.com

STYLE Distilled Gin

ORIGIN Washington D.C. – U.S.

BOTANICALS 12. Angelica, cassia, celery seeds, coriander, fennel, grains of paradise, grapefruit, juniper, lemon, lemongrass, orris root, sage.

PRODUCTION Distilled at New Columbia Distillers on the outskirts of Washington D.C., from a base of New Make Soft Red Winter Wheat Spirit.

Each single batch is handcrafted within a month. The base is a wheat distillate made in a traditional copper alembic. In the second distillation, botanicals are placed in a basket through which the distillation steam passes. Next is filtration, then bottling by hand.

NOTES The name Green Hat refers to historical figure George Cassiday who, having returned from the First World War, took to the production and sale of alcohol in Washington during the Prohibition years. As it happened, Cassiday, known to wear a green hat, soon began to find many clients among deputies and senators. For this reason, although he was arrested several times, Cassiday never spent a single night in prison. Around 1930, using the pseudonym Cassiday, he published some articles highlighting the hypocrisy of American politicians of the time, whom he labelled Prohibitionists in word, but not in deed.

New Columbia Distillers is a family business owned by Michael Lowe and his wife, Melissa Kroning. He is a former lawyer, passionate about cocktails and spirits, a passion that led him to leave his profession and embark on this new path. When the distillery opened in 2011, it was the capital's first artisanal establishment of its kind.

RELEASE Green Hat Navy Strength: 57% abv.

Green Hat Spring/Summer: 44.8% abv. With added rosemary, cherry blossom and lemon.

Green Hat Fall / Winter Ginavit: 45.2% abv. A blend of the classic juniper, a caraway acquavite and a portion of rye. There is an XO version that is barrel aged for 15 months then bottled at 46% abv.

G&T GARNISH Lemongrass, sage leaf, yellow grapefruit peel.

COCKTAILS Claridge, Alaska, Gin Cobbler.

HALF HITCH DISTILLED GIN
40% ABV

WEB www.halfhitch.london

STYLE Distilled Compound Gin

ORIGIN Camden Lock, London – United Kingdom

BOTANICALS 13. Juniper, black tea from Malawi, Calabrian bergamot, pepper, wood, hay, angelica root, cassia bark, coriander seeds, liquorice root, sweet lemon, orange peel, iris root.

PRODUCTION The base distillate comes from the well-known Langley distillery, to which Mark Holdswort adds distillates and tinctures. Beginning with a basic gin, a London Dry Gin purchased externally, composed of a 100% British wheat base, juniper, coriander, cassia, liquorice, angelica root, lemon and orange peel, and orris root. To this, highest quality tinctures of black tea and bergamot are added, along with infusions of English wood and pepper. Finally, Holdswort adds a hay distillate made via the vacuum technique.

NOTES Launched in 2015, production takes place in the former Camden Lock warehouses. From 1869 to 1964 in the nearby Clerkenwell area, many famous brands were once found here, such as Gordon's and Booths. In fact, production capacity in Camden distilleries once exceeded 3 million liters in a year (1897). At that time, one train a week would depart from Locks carrying exclusively gin destined for other parts of the world. From this time, unfortunately there remain very few memories and some scant references, such as an area street bearing the name "Juniper Crescent". At the end of 2014, however, Mark Holdsworth (formerly of Bacardi) launched Half Hitch Gin, taking part of the process back to Camden and recovering a wonderful old abandoned Victorian building, "The Interchange". The building also features a canal known locally as "Dead Dog Hole", where 6 barges could moor.

A "half hitch" was the type of rope knot used to moor boats along Camden Lock in the industrial era.

G&T GARNISH Garda lemon zest, sprig of Greek hay, bergamot peel.

COCKTAILS Bronx, Old Fashioned, Julep.

HENDRICK'S GIN
41.4% ABV

WEB www.hendricksgin.com

STYLE Distilled Gin

ORIGIN Girvan, Ayrshire – Scotland

BOTANICALS 14. Juniper, coriander, angelica root, orris root, orange and lemon peel, cubebs, grains of paradise, elderberry, chamomile, yarrow, cumin seeds plus essence of damask rose, Dutch cucumber.

PRODUCTION In 1999, William Grant & Sons decided to make a Premium gin inspired by traditional English gardens, distilled in two stills: a John Dore Carter Head from 1948 and an 1860 small Bennett pot still. Botanicals are worked separately in the two stills. In the Bennett, they are infused for 24 hours before distillation, while in the Carter Head they are placed in the dedicated basket for contact with the steam. The results of the two stills are then combined, to create a homogeneous sensory profile. At this point, before diluting with water to lower alcohol content, essences of cucumber and damask rose are added.

NOTES Launched in 2000 in the United States, this is a brand that in my opinion inspired a landmark shift in the Gin & Tonic, both in production and style.

The label bears 1886, a reference to the year William Grant opened his distillery.

The recipe seems to derive from Grant's inspiration after taking tea in the garden, with cucumber sandwiches and Gin & Tonic. This pleasant combination he then transposed onto his idea of gin, which he then dedicated to a local gardener, who was named "Hendrick".

RELEASE Hendrick's Kanaracuni: 44% abv. In 2013, Lesley Gracie and Global Ambassador David Piper ventured into the forests of Venezuela, accompanied by an explorer and a botanist. Lesley armed herself with a 10-liter pot still, wanting to distil on the spot any rare botanicals they might come across, including the well-known "Scorpion Tail". While distilling in the heart of the jungle, she managed to obtain 8.4 liters of what is now this gin's key ingredient, called "Kanaracuni" after the name of their supporting village. The base is the classic Hendrick's, only with different proportions from the two stills, with the addition of the two essences. Annual production is quite small: only 560 bottles.

Hendrick's Orbium: 43.4% abv. Classic Hendrick's with added quinine, wormwood and blue lotus flower.

Midsummer Solstice: 43.4% abv. A Hendrick's Gin with an accentuated floral blend.

G&T GARNISH Lemon peel, cucumber peel or slice, rosebud or fresh chamomile.

COCKTAILS Corpse Reviver No. 2, Cardinale, Aviation.

HERNÖ GIN
40.5% ABV

WEB www.herno.se

STYLE London Dry Gin

ORIGIN Hillgren – Sweden

BOTANICALS 8. Juniper, coriander, meadowsweet, cassia bark, black pepper, vanilla, Nordic cranberries, lemon zest.

PRODUCTION The distillery is located in Dala, a village near Härnösand, Sweden. The gin was launched on the market on December 1, 2012. The master distiller and founding partner is Jon Hillgren (graduate of The Institute of Brewing and Distilling in London). The still used is a 250-liter copper pot still named "Kierstin". The basic distillate is neutral, made from local wheat. The botanicals are all natural and organic, harvested exclusively in Sweden. The processing style is typical of a London Dry Gin. In this case, the spirit is macerated with juniper and coriander for about 18 hours. Next, other botanicals are added and everything is distilled. The water used to lower the alcohol content is taken from the producers' personal well.

NOTES Given the presence of meadowsweet, this excellent product cannot be marketed in the United States. To avoid losing such an important market, the recipe has been slightly modified by replacing meadowsweet with yarrow.

RELEASE Hernö Navy Strength Gin: 57% abv. Same recipe as the classic.

Hernö Old Tom Gin: 43% abv. Made in September 2014. Its sweetness owes to a greater amount of meadowsweet, as well as the addition of honey.

Hernö Juniper Cask Aged Gin: 47% abv. Matured for 30 days in a 39.25-liter juniper cask built specifically for this production.

Hernö Black currant Gin: 28% abv. Only 588 bottles, presented in the summer of 2013, and made with locally harvested black currants.

G&T GARNISH Lemon peel, small red berries, a light dusting of Tonka bean.

COCKTAILS Gin Spicy Fifty, Gin Crusta, Silver Fizz.

ISH GIN
41% ABV

WEB www.ishgin.com

STYLE London Dry Gin

ORIGIN London – United Kingdom

BOTANICALS 12. Juniper, coriander seeds, angelica, almond, orris root, nutmeg, cinnamon, cassia, liquorice, lemon and orange peel, winter savory (and possibly ginger).

PRODUCTION Designed by Ellen Baker, this is a Spanish-inspired gin that references her beloved "Bristol bar" street in Madrid. Distilled 5 times in a traditional pot still at the Thames Distillery in the heart of London. Macerated for 24 hours before being distilled.

NOTES The 3 letters in the name are an acronym for Irresistible Scandalous Hallmark. Marked by a high amount of juniper.

RELEASE ISH Limed Distilled Gin 40% abv. A very marked note of lime.

G&T GARNISH Lemon peel, orange peel, fresh ginger slice.

COCKTAILS Bronx, Park Avenue Cocktail, Gin Sour.

JENSEN'S BERMONDSEY DRY GIN
43% ABV

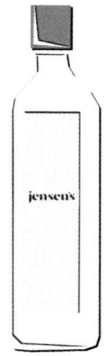

WEB www.bermondseygin.com

STYLE London Dry Gin

ORIGIN London – United Kingdom

BOTANICALS Unknown. Only juniper is stated, but I believe this gin uses only 19th-century botanicals like coriander, juniper, iris, angelica, almond, liquorice, orange and lemon peel.

PRODUCTION Distilled at the Bermondsey Distillery, founded by Danishman Christian Jensen who, having come from a completely different background (finance and banks), but who always loved the world of spirits and cocktails, decided to lean on one of the world's leading experts: Charles Maxwell of Thames Distillery. The first production was 1,500 bottles, with which Jensen immediately began to promote his product in London, seeing excellent results. After almost 10 years of production at Thames, in 2013 Jensen established his own small distillery, the Bermondsey Distillery under the Maltby Street Railway Arches in South London (near London Bridge). Here he installed a wonderful John Dore still, a replica of the famous Tom Thumb still at Thames. A curiosity: Jensen's right arm is Anne Brock, a female master distiller.

NOTES When the first production was ready, the boxes delivered by Thames were so many that Jensen was forced to store some in his apartment.

Great attention is paid to the absolute freshness and quality of every single botanical element used. All citrus fruits are peeled exclusively by hand.

RELEASE Jensen's Old Tom Gin: 43% abv. Made from an ancient recipe (1840) and with no added sugars. Instead, it is sweetened by the softness of spices such as liquorice and (likely) also eucalyptus.

G&T GARNISH Thin lime slice, lemon peel, fresh mint bunch.

COCKTAILS Pegu Club, Negroni, Marguerite Cocktail.

JINZU GIN
41.3% ABV

WEB www.diageo.com

STYLE Distilled Compound Gin

ORIGIN Scotland – United Kingdom

BOTANICALS 5+1. Juniper, coriander, angelica, cherry blossom, yuzu, sake.

PRODUCTION Distilled at Diageo's Cameronbridge Gin Distillery, this is a wonderful East meets West blend, featuring traditional botanicals combined with cherry blossoms, yuzu and sake. Born from the union of a juniper distillate and a sake distillate, with the addition of essences and a final distillation. Supervising everything until 2015 was Tom Nichol of Tanqueray Distillery (an icon among master distillers), who retired that year.

Juniper, coriander, and angelica are infused into a neutral grain spirit in a traditional copper pot still and steeped for a short time before the cherry blossoms and yuzu are added. After being released at 82% abv, the distillate is enriched with Junmai quality sake and brought to 60% abv. The final phase involves a slow dilution with demineralized Scottish water until lowered to 41.3% abv. About 1,600 bottles are produced from each distillation.

NOTES The bottle was designed by bartender Dee Davies who won the Diageo's Show Your Spirit Competition in 2103.

The name of this gin is an homage to the Japanese river that flows through Toyama, a city on the coast of the Sea of Japan. Thousands of cherry trees grow along its banks.

Cameronbridge is instead the name of the bridge over the river Leven, which runs right past the distillery, which was previously owned by John Haig (starting in 1824), a cousin of Robert Stein, famous for his invention known as the continuous still.

G&T GARNISH Green apple or candied cherry blossom.

COCKTAILS French 75, Singapore Sling, Gin Sour.

JUNIPERO GIN
49.3% ABV

WEB www.anchordistilling.com – www.hotalingandco.com

STYLE London Dry Gin

ORIGIN Potrero Hill, San Francisco, California – U.S.

BOTANICALS 12 stated. Juniper, coriander, liquorice, angelica, anise, cardamom, cassia, coriander, cubeb berries, grains of paradise, lemon and orange peel, iris root.

PRODUCTION Founder Fritz Maytag opened Anchor Distilling, now known as Hotalin & Co and a subsidiary of the Anchor Brewing Company (famous for Anchor Beer), in 1993 as one of the first American craft distilleries. Junipero was launched in 1998. For distillation, they use a tiny copper pot still.

NOTES The brand is named for Saint Junìpero Serra, a Franciscan friar and an important historical figure in the history of San Francisco.

RELEASE Anchor Gin Old Tom: 45% abv. Softened with liquorice, stevia and star anise.

G&T GARNISH Lemon peel, star anise, yellow grapefruit peel.

COCKTAILS Gimlet, Silver Bulleit, American John Collins.

KI NO BI KYOTO DRY GIN
45.7% ABV

WEB www.kyotodistillery.jp

STYLE Distilled Compound Gin

ORIGIN Kyoto – Japan

BOTANICALS 11. Juniper, iris, yellow yuzu, bamboo leaves, Gyokuro green tea, Sansho pepper, red shiso leaves, Japanese cypress (hinoki), ginger, lemon, Kinomi pepper leaves.

PRODUCTION Alex Davies is the master distiller from the Cotswolds Distillery. The botanicals are divided into 6 categories (basic, citrus, tea, herbs, spices and flowers) distilled separately and then assembled. The distillery has two 140-liter discontinuous stills, with an integrated botanical basket and a 450-liter swan neck. The base for the maceration is a neutral rice distillate. The water used to dilute the alcohol content comes from Fushimi.

NOTES The Kyoto Distillery is the first artisanal distillery in Japan.

Ki No Bi means "The beauty of the seasons".

The karakami, a traditional Japanese art form involving decorative paper, colored and printed with designs, are produced by Kira Karacho, the oldest atelier specializing in this art, founded in Kyoto in 1624.

For this gin, the karakami are made on washi paper using around 600 wooden blocks handed down from generation to generation.

RELEASE Ki No Bi Ginza's Mori: 45.7% abv. An homage to the legendary Japanese bartender Takao Mori, bringing together 6 spirits obtained from 11 varieties of plants.

Ki No Bi KI No Tea: 45.7% abv. The result of a collaboration with tea assembler and producer, Hori-Shichimeien, whose family has owned tea plantations since 1879 in the Uji region south of Kyoto. Includes Sencha and Gyokuro tea, yuzu, lemon, juniper, iris and hinoki.

Ki No Bi Navy Strength, called "Sei": 54.5% abv. Made with 11 botanicals.

Ki No Tuo: 47% abv. Old Tom style sweetened with kokuto (also known as "black sugar") from the island of Okinawa. The two kanji characters stand for "island" and "sugar".

Ki No Bi Cask Aged: 48% abv. In collaboration with Kamiasobi Noh Troupe, and matured in sherry barrels previously used for Karuizawa Single Malt Whiskey. Today there are 6 different aged editions, among which the #5 Caroni Cask for the Maison du Whiskey stands out as notable.

G&T GARNISH Yuzu peel, shiso leaves, thin slice of ginger.

COCKTAILS Gin Sling, Clover Club, Gimlet.

LE TRIBUTE GIN
43% ABV

WEB www.letribute.com

STYLE Distilled Compound Gin

ORIGIN Vilanova – Spain

BOTANICALS 11. Juniper, lime, kumquat, pink grapefruit, green grapefruit, mandarin, cardamom, sweet and bitter oranges, lemons, lemongrass.

PRODUCTION Destilerias MG, which produce 7 different distillates from single macerations in neutral alcohol of barley and wheat: 1) juniper, 2) lime, 3) kumquat, 4) pink grapefruit and green grapefruit, 5) mandarin, 6) cardamom, sweet oranges and bitter and lemons, 7) lemongrass.

Each individual distillate is then blended, and the alcohol content lowered.

NOTES This gin features a 100% copper cap. It is the result of two years of work.

Destilerias MG, former producers of MG Gin and Gin Mare, are a large family-run business in Vilanova, near Barcelona, and their production is decidedly prominent. Founded by Giro in 1835, this is one of Spain's oldest distilleries. The Le Tribute brand is dedicated to family and tradition, and they boast 5 generations. (Formerly they were manufacturers of medicines, which explains the bottle's features, an homage or, better put, a tribute to the world of gin).

The juniper is harvested by hand at the Giro family's farmhouse in Teruel.

G&T GARNISH Pink grapefruit peel, lemon peel, piece of lemongrass.

COCKTAILS Corpse Reviver No. 2, Clover Club, Gin Smash.

LEOPOLD'S GIN
40% ABV

WEB www.leopoldbros.com

STYLE Distilled Compound Gin

ORIGIN Denver, Colorado – U.S.

BOTANICALS 6 stated. Juniper, coriander, orange and pomelo peel, orris root, cardamom.

PRODUCTION Launched in 1999, distilled since 2001 at Leopold Bro iin small batch copper pot still. Each single botanical is distilled separately and each production consists of just 50 cases. A curiosity: the base distillate is made from a mix of potatoes, barley malt and wheat.

NOTES Each bottle is numbered by hand.

RELEASE Leopold's Navy Strength American Gin: 57% abv. With added bergamot.

Leopold's Summer Gin: 47% abv. Separate distillation of juniper berries, coriander, blood oranges, lemon myrtle leaves and helichrysum (also known as the "immortal flower").

G&T GARNISH Pomelo zest, lemon peel.

COCKTAILS White Lady, Gin Fizz, Ramos Fizz.

Nº3 LONDON DRY GIN
46% ABV

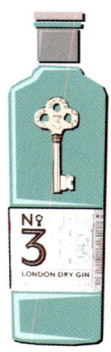

WEB www.no3gin.com

STYLE London Dry Gin

ORIGIN Schiedam – The Netherlands

BOTANICALS 6. Italian juniper, coriander from Morocco, orange peel, angelica root, cardamom, grapefruit peel.

PRODUCTION Distilled at De Kuyper Royal Distillers yet designed by a legendary London shop with a 300-year history that wished to work with a company of equally long tradition in the distillation field. Every batch is overseen by distiller David Clutton, who jealously preserves the recipe of this very classic London Dry Gin. The spices are weighed and left to macerate in alcohol for a day before being gently distilled in a copper pot still, where they are blended and left to rest overnight.

NOTES The name is a dedication to the St James's Street building where Berry Bros. & Rudd, Britain's oldest wine and spirits merchant, is based.

Number 3 is no accidental label but rather the street address, and moreover indicates the 3 different spices and the 3 fruits that constitute the botanical blend.

The key pictured on the bottle of this gin (born in 2010) is a replica of the one used to open The Parlor room, one of the most secret and ancient rooms in the entire shop, there since 1668. According to legend, some of the most precious liquid treasures ever owned by the world's most "powerful" are still housed inside the room today: from priceless cognacs to unique wines (more unique than rare). The key also symbolizes trust.

The bottle itself is a tribute to the history of this distillate, recalling a time when it was still transported in wooden boxes and the bottles were shaped accordingly to fit snugly inside the boxes, preventing them from moving about and consequently breaking. The shape, called an open pontil, dates to the 17th century. And even the color, in addition to recalling the greenness of England green, is the same as the color the barrels once used, if a slightly darker shade.

G&T GARNISH Garda lemon or citron zest.

COCKTAILS Classic Martini, Gin Fizz, Negroni.

LUZ FASHIONABLE GIN
45% ABV

WEB www.luz-gin.it

STYLE Italian Distilled Compound Gin

ORIGIN Riva del Garda (Trento) – Italy

BOTANICALS 9. Juniper, laurel, olive tree, woodruff, rosemary, sage, clary sage, common mint, lemons (all botanicals originating exclusively from the Lake Garda and Trentino area).

PRODUCTION Production launched in 2013, after several years of experiments and technical tests to obtain the most representative expression of Lake Garda, where most of the botanicals used to make this lovely gin are gathered. Initially distilled in Alto Adige (South Tyrol). At that time, however, creator Leonardo Veronesi was not entirely satisfied with the results, so he launched an exhausting search, hoping to bring about the highest qualitative expression to his product, which he found and consolidated at the Marzadro Distilleries, on 1 February 2018. Distillation takes place in 1,000-liter copper stills in a bain marie discontinuous system. The infusions are separated according to type and origin of the botanicals, using different alcohol levels and, for a minimum of 12 / maximum 96 hours. The base alcohol is made from grains, while the water used to lower the alcohol content comes from the springs of Monte Baldo in Trentino. To preserve every precious element, no filtration of any kind is carried out.

NOTES Leonardo Veronesi is a well-known, highly competent barman in Italy. He owns the nationally popular Rivabar in Riva del Garda (www.rivabar.it).

G&T GARNISH Garda lemon peel, clary sage leaf, olive branch.

COCKTAILS Gin Sour, Gin Sling al Sambuco, Rosemary Fizz.

MACARONESIAN GIN
40% ABV

WEB www.macaronesiangin.com

STYLE Distilled Gin

ORIGIN Canary Islands – Spain

BOTANICALS 6 stated. Juniper, cardamom, angelica root, liquorice, lemon peel, orange peel.

PRODUCTION Produced in the Canary Islands by DSC (Distileria Santa Cruz) in San Miguel de Abona, using botanicals grown and harvested exclusively in the production area, an area blessed with a wonderful subtropical oceanic climate that assures constant mild temperatures. Dilution for lowering of the alcoholic content uses very pure water, naturally filtered using islan volcanic rocks.

NOTES The name derives from the Greek μακαρων νεσοι (makàron nêsoi) meaning "Islands of the blessed", an expression used by Hellenic geographers to describe the beautiful islands found beyond Strait of Gibraltar: Canaries, Azores, Madeira, Cape Verde and Islas Salvajes - Wild Islands. They were also called "Fortunate Islands" on account of a belief that the gods welcomed mortals with extraordinary gifts here.

The bottle is produced in Germany, made of organic clay also sourced locally, and bleached using natural colors, then fired at temps above 1200 ° C.

RELEASE Macaronesian Gin Eternal Spring Strawberry Flavour: 37.5% abv.

Macaronesian Brownie Cream: 40% abv.

G&T GARNISH Citron peel, liquorice root slice, lemongrass stick.

COCKTAILS Florodora, Mayfair Cocktail, Gin Sling.

GIN MARE
42.7% ABV

WEB www.ginmare.com

STYLE Distilled Compound Gin

ORIGIN Vilanova i la Geltrù, Barcelona – Spain

BOTANICALS 8 (actually 9). Spanish juniper, Spanish Arbequine olives, cardamom, coriander, thyme from Turkey, basil from Italy, rosemary from Greece, citrus fruits (bitter orange from Valencia, sweet orange from Seville, lemons from Lleida).

PRODUCTION The small batch pot still used in production is located inside a chapel where fishermen would go before setting sail. The original was destroyed by a storm.

Every botanical is macerated individually in 200 liters of neutral grain alcohol, then water is added to lower the alcohol content to 50% abv. The actual distillation takes 90 minutes.

Each botanical is left to macerate directly inside the still for 24-36 hours.

A liter of distillate requires 5 kg of olives.

NOTES After working with wine and spirits since 1835, the Ribot family started working with gin in the 1940s. It was Manuel Giro Sr who launched MG Gin, still one of the best-selling gins in Spain today.

Gin Mare was launched by grandsons Mark and Manuel Jr, who wished to create a new type of gin inspired by the Mediterranean climate and using locally sourced botanicals. This new gin, made in collaboration with Global Premium Brands, dates to 2007.

They started by distilling 45 botanicals separately, and in two years of research they conducted 90 trials. Up to 15 kg of olives are used for each distillation batch, which must pass through a crusher before being pitted.

Citrus fruits are macerated for a whole year in a neutral spirit of 50% abv. Each year the distillery uses about 200 kg of orange peel and about 80 kg of lemon peel.

Most other botanicals are macerated separately for about 36 hours and then individually distilled in a 250 liter Florentine Still. The distillation takes about 4 hours.

Inside the chapel there is a Latin motto that says: "Mundus appellatur caelum, terra et mare" (The world is defined in sky, earth and sea).

G&T GARNISH Sprig of fresh rosemary, basil leaves, Garda lemon peel.

COCKTAILS Dirty Martini, Red Snapper, Gin Smash.

MAYFAIR GIN
40% ABV

WEB www.mayfairbrands.com

STYLE London Dry Gin

ORIGIN Warrington, Cheshire – United Kingdom

BOTANICALS 5 stated. Juniper, coriander, angelica, iris root powder, winter savory.

PRODUCTION Produced by Thames Distillery and expertly managed by Charles Maxwell using their historic stills: Christened, Thumbelina and Tom Thumb. They don't readily reveal their secrets. Yet, if the base of this elegant gin is their vodka, there will be no doubting about the quality of their final product, sure to be of the highest level. Distillation starts with a base made from top quality wheat grown in the United Kingdom, and is carried out in a double column still 6 times. The final stage, seventh redistillation of all botanicals, takes place in a copper pot still named "Tom Thumb".

NOTES Owned by Mayfair Brands Company.

Named for one of the most exclusive districts in London, Mayfair is located near Hyde Park, Regent Street and Piccadilly. The area is famous for its unparalleled luxury along with numerous embassies, residences of famous personalities and 5-star hotels.

RELEASE Mayfair Vodka: 40% abv.

G&T GARNISH Lemon or grapefruit peel.

COCKTAILS Martini Cocktail, Cardinale, Bijou.

MOMBASA CLUB DRY GIN
41.5% ABV

WEB www.mombasagin.com

STYLE London Dry Gin

ORIGIN London – United Kingdom

BOTANICALS None stated. Juniper, cassia bark, coriander, cloves, angelica, citrus, cinnamon, liquorice.

PRODUCTION Owned by Unesdi Distribuciones S.A. of El Puerto de Santa Maria (Cadiz, Spain), but distilled at the famous Thames Distillery in the heart of London using a traditional small batch method and a neutral alcohol distilled 4 times, then combined with a careful selection of natural ingredients, immersed directly in the spirit. After distillation, this gin is shipped to Spain, where it is diluted with local waters.

NOTES The Mombasa Club Gin has distant, Victorian-era origins. At the end of the 19th century, during the colonial war to conquer Africa, the powerful British Empire took hold of Mombasa in Kenya, and transformed it into the central port and commercial center for the whole of East Africa. The Mombasa Club was born here, founded in 1885 by the British East Africa Company. The club was reserved for British East Africa Protectorate officers, those born in the United Kingdom and who held official positions or were in the employ of Imperial East Africa Company. Inside, one could taste the Mombasa Club Gin, produced in England exclusively for club members, for whom the club quickly became a familiar, welcoming, almost idyllic locale for spending one's leisure time, whether participating in lively debates or exchanging news while enjoying a Mombasa Club Gin & Tonic. Today Mombasa Club Gin continues to represent that period of British colonialism and those who were part of it.

RELEASE Mombasa Club Colonel's Reserve: 43.5% abv. 11th Limited Edition. Botanicals—cumin, calamus (or sweet flag) and coriander seeds in place of angelica—are macerated overnight. Distilled in an original John Dore.

Mombasa Club Strawberry Edition: 37.5% abv. Distilled 3 times, flavored with strawberries.

G&T GARNISH Lemon peel, sprig of calamint, star anise.

COCKTAILS White Lady, Alaska, Gin Cobbler.

MONKEY 47 SCHWARZWALD DRY GIN
47% ABV

WEB www.monkey47.com

STYLE London Dry Gin

ORIGIN Wolfach, Black Forest – Germany

BOTANICALS 47, including juniper, angelica root, acacia blossom, chamomile, cinnamon, cardamom, cassia, cloves, coriander, lavender, ginger, liquorice, elderflower, almond, kaffir lime, abelmosk (or musk mallow), Chinese hibiscus, calamus (or sweet flag), bitter orange, blackberry, lemon verbena, cranberry, hawthorn, honeysuckle, jasmine, lemon and pomelo peel, nutmeg, iris, rose hips, blackthorn, sage, lemon balm, lemongrass, bergamot, grains of paradise, allspice, cubeb berries, allspice, citron and rosehip.

From the Black Forest: blackberry leaves, cranberries, spruce sprouts.

PRODUCTION The Black Forest Distillery launched production in 2008. The "hard" spices used are macerated directly in molasses alcohol and distilled in France for 36 hours, while the "delicate" ones are processed exclusively via vapor infusion. The distillate then rests in special terracotta containers, and the alcohol content is lowered with water sourced in the Black Forest.

NOTES In July 1945, Montgomery Collins, an officer of the Royal Air Force, was stationed in Berlin, where he adopted a little monkey named "Max" from a local zoo. Collins left the RAF in 1951 and moved to the Black Forest, where he opened The Wild Monkey guesthouse in Max's honor. Relying on the great variety of herbs and spices in the area, along with local expertise in the art of distillation, Collins started making a gin that became a distinctive aspect of his guesthouse. Alexander Stein, who comes from a family of distillers, rediscovered the recipe.

Each year they create a special Distillers Cut, which features an additional, more dominant botanical element. It matures for 12 months in terracotta containers. Each edition produces around 4,000 bottles.

Monkey 47 was launched in 2010.

RELEASE Monkey 47 Distillers Cut: 47% abv.

2011: Hibiscus, 2012: Maiwipferl, 2013/14: Oxalis triangularis, 2015: Meum Athamanticum, 2016: Abietes Melle, 2017: Iva, 2018: Red Mustard Cress, 2019: Myristicae Arillus (macis)

Monkey 47 Sloe Gin: 29% abv. With wild sloe berries.

Monkey 47 Barrel Cut: 47% abv. Made in new, handmade, toasted mulberry wood barrels.

G&T GARNISH Lemon verbena leaves, fresh chamomile flowers, pink grapefruit peel.

COCKTAILS Bee's Knees, Pegu Club Cocktail, Gin Sling.

Nº209 GIN
46% ABV

WEB www.distillery209.com

STYLE Distilled Gin

ORIGIN San Francisco – U.S.

BOTANICALS 11, including juniper from Italy, cassia bark from Indonesia, cardamom seeds from Guatemala, lemon peel from Spain, coriander seeds from Romania, angelica root from England and bergamot peels from Calabria (possibly also iris and liquorice).

PRODUCTION Released in 2005, this gin is distilled in a pot still built in Scotland. Botanicals are infused overnight in a neutral alcohol spirit. After distillation, which lasts about 11 hours, water is used that comes from the Sierra Nevada mountains. The alcohol base is distilled 5 times. The wheat comes from the Midwest. The still, inspired by those of Glenmorangie, was built by Forsythe's and can contain almost 400 liters of distillate.

NOTES In 1870, William Scheffler of New York acquired the patent rights of a new design of still in California. In 1880, he purchased the Edge Hill Estate in St. Helena, Napa Valley, one of the most important wineries at the time. Two years later, he built a brick and stone distillery and registered it with the federal government under license number 209, which is the number painted above the entrance. During Prohibition, the distillery closed and production ceased.

In 1999, Leslie Rudd took over as new manager of the Edge Hill property, which he purchased for his vineyards. One day, while visiting the estate, he was passing by the barn and noticed the inscription, Registered Distillery No. 209, painted above the entrance door. From there the idea of bringing the old Edge Hill factory back to life came quickly. He constructed a new distillery at Pier 50 in San Francisco and began distilling... practically right on the sea!

RELEASE No. 209 Barrel Reserve Gin: 46% abv. Aged in French barrels that previously contained California sauvignon blanc.

No. 209 Barrel Reserve Gin: 46% abv. Produced in ex-chardonnay barrels.

No. 209 Barrel Reserve Gin: 46% abv. Produced in ex-cabernet sauvignon barrels.

No. 209 Kosher-for-passover-Gin: 46% abv. With the addition of California laurel from Napa Valley.

G&T GARNISH Lemon or bergamot peel.

COCKTAILS Gin Toddy, Last Word, Gin Julep.

NOLET'S DRY GIN
47.6% ABV

WEB www.noletsgin.com

STYLE Distilled Compound Gin

ORIGIN Schiedam – The Netherlands

BOTANICALS 7. Juniper, lime, orris root, liquorice, peaches, raspberries, damask rose.

PRODUCTION Produced and distilled at the Nolet Distillery in Schiedam, on the banks of the Buitenhaven canal, from a recipe created by Carolus Nolet. Distillation is divided into two significant operations, and takes place in a 300-liter copper pot still, which has been flanked by a long rectification column to allow for greater alcohol extraction at 90% abv. To begin, a true London Dry Gin is made from a neutral wheat alcohol produced entirely in the distillery (this is one of the few distilleries in the world that oversees the entire production chain). To this, juniper, lime, liquorice and iris root are added. Then, the individual macerations of peach, rose and raspberries are carried out in 50% abv alcohol at 50 ° C for 24 hours. The botanicals will then be distilled and combined with the first London Dry Gin.

NOTES Founded in 1691 by Joannes Nolet, this is the oldest existing distillery in The Netherlands, and likely one of the oldest in Europe, if not in the entire world.

RELEASE Nolet's Dry Gin The Reserve: 52.3% abv. A limited edition produced directly by Carolus Nolet, tenth in the generational line of the family. Characterized by verbena and saffron, today this gin is one of the most expensive in the world, with a minimum price of 600 euros (approximately 700 US dollars) per bottle. It is numbered and signed by Nolet.

Ketel 1 Jenever: 35% abv. A completely ambachtelijke product (meaning "artisanal" in Danish) with only 3% of malt wine (made by Filliers), expertly blended with wheat, corn and rye. The malt wine ages from 12 to 18 months at the distillery in 220-liter French wood barrels, and is then redistilled with 14 secret botanicals in Still No. 1 or No. 7 (No. 1 is the oldest in the distillery, built in 1853).

Ketel 1 Matuur: 38.4% abv. A blend of neutral wheat alcohol and a blend of malt wines. Aged for at least 8 years in new American oak barrels. The malt wine is partially redistilled according to a secret recipe.

G&T GARNISH Peach slice, Amalfi lemon peel, fresh raspberry.

COCKTAILS Mayfair Cocktail, Clover Club, Three to One Cocktail.

NORDES ATLANTIC GALICIAN GIN
40% ABV

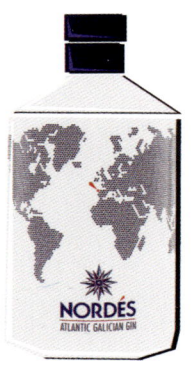

WEB www.nordesgin.com

STYLE Likely a London Dry Gin

ORIGIN San Pedro de Sarandón, Santiago de Compostela – Spain

BOTANICALS 11. Six wild Galician botanicals: sage, bay leaf, verbena herb, eucalyptus, peppermint, glasswort.

Five overseas botanicals: juniper, ginger, cardamom, hibiscus flowers, black tea.

PRODUCTION The base distillate is produced using Galician albariño grapes. Distillation takes place via the amodiño method ("slowly") in order preserve the aromatic elements.

NOTES This gin's story started in a restaurant, with three friends chatting over a plate of octopus. One is a famous sommelier who won the Nariz de Oro prize (a prestigious Spanish wine competition) in 2004, the other is a Galician wine entrepreneur, and the third is a skilled master distiller. Their dream was to produce a true Galician gin, using the best ingredients available in their land.

The gin takes its name from the north wind, which delivers freshness from the Atlantic Ocean.

The characteristic white bottle of Nordés is inspired by the traditional ceramics of Sargadelos, and is produced by Galician artisans.

A compass and a world map appear on the bottle, highlighting the Galician origin of the product.

The Ulla river runs through the production area.

RELEASE Nordes Atlantic Galician Vodka: 40% abv. Distilled 3 times from an alcohol base obtained from potatoes.

G&T GARNISH Sprig of fresh mint, citron peel, slice of ginger.

COCKTAILS Southside, Gin Fix, Gin Julep.

OLD ENGLISH GIN
44% ABV

WEB www.oldenglishgin.com

STYLE Old Tom Gin – Distilled Gin

ORIGIN Denmark, but also produced in Birmingham – United Kingdom

BOTANICALS 10+1 secret. Juniper, coriander, lemon and orange peel, cassia, angelica, liquorice, iris, cinnamon, nutmeg.

PRODUCTION Owned by Hammer & Son LTD.

Created by Henrick Hammer, member and recognized judge of the IWSC (International Wine & Spirit Competition, founded in 1969 in London), this gin is distilled at Langleys Distillery in Birmingham. The processing style is that of a London Dry Gin, using a pre-infusion of the botanicals in wheat alcohol, and a subsequent distillation in a copper pot still called "Angela".

White cane sugar syrup equal to 4% grams per liter is added.

NOTES It was inspired by a recipe from 1783.

Sold in champagne bottles, which are much more resistant, as the producers have decided to follow an ancient system of reusing containers.

The sealing wax used is organic and of natural origin, while the printing uses eco-friendly ink.

If you take a close look at the bottle, you might see a hidden clue regarding this gin's secret ingredient. Above the shield picturing the two crossed hammers is a cross-sectional drawing of a cardamom seed. Could this be the undeclared ingredient?

The bottle also features hammers bearing 4 initials, which could be a further clue about Henrik's father, an important figure in his own right. He was called Hudi after the legendary King of the Britons Rud Hud Hudibras, who had 4 children. Each hammer could be the initials of each of them.

G&T GARNISH Garda lemon peel, bergamot peel, chamomile flowers.

COCKTAILS Ford Cocktail, Tuxedo, Turf.

O'NDINA GIN
45% ABV

WEB www.camparigroup.com

STYLE Distilled Compound Gin

ORIGIN Novi Ligure (Alessandria) – Italy

BOTANICALS 19 stated. Juniper, coriander, fennel, cardamom, Sichuan pepper, liquorice, thyme, marjoram, sage, cinnamon, nutmeg, galangal, basil, 100% Italian citrus fruits from various sources.

PRODUCTION Designed and produced entirely in Italy in the Novi Ligure distillery. Overseen by master herbalist and distiller Bruno Malavasi, who wanted to create a type of Mediterranean gin using such key botanicals as the emblematically Italian Genovese basil. Collected fresh and frozen on site, the basil is immediately macerated in grain alcohol for a few days, then distilled in a vacuum still. The remaining 18 botanicals are carefully dosed, infused and distilled all together, as if a full-fledged London Dry Gin.

Finally, the two distillates are assembled in secret proportions and left to rest for a few weeks in stainless steel tanks.

NOTES The name is a dedication to a great love story.

G&T GARNISH Sprig of fresh basil, Amalfi coast lemon peel, sprig of fresh thyme.

COCKTAILS Mediterranean Martini, Dirty Martini, Red Snapper, Basil Gin Smash.

OPIHR ORIENTAL SPICED GIN
40% ABV

WEB www.opihr.com

STYLE London Dry Gin

ORIGIN Warrington, Cheshire – United Kingdom

BOTANICALS 12 non stated. Juniper from Italy, cubeb berries from Malaysia, black Tellicherry pepper, cardamom and ginger from Malabar (India), cumin from Turkey, coriander seeds from Morocco, bitter oranges from Spain, angelica from Germany, cassia from China, grapefruit peel, and curry (according to rumors).

PRODUCTION Launched in October 2013 and produced by Greenall's Distillery (Quintessential Brands Group), this gin was created by the amazing master distiller Joanne Simcock Moore, the seventh master distiller in the distillery's 250-year history and one of the world's few female master distillers. Distillation takes place in a traditional small batch copper pot still filled with neutral two-distillation wheat alcohol. The botanicals are left to macerate directly for 24 hours before the third distillation.

NOTES The name derives from the legendary port of Ophir, where, according to the Bible, every 3 years King Solomon would receive a shipment of gold, silver, sandalwood, precious stones, ivory, monkeys and peacocks from Africa.

Tomé Lopes, Vasco da Gama's companion, hypothesized that Ophir may have been the ancient name of Great Zimbabwe. Others have suggested it could be placed in Sofala, Mozambique, while still others have linked it to the port of Ezion Geber on the Red Sea (Gulf of Aqaba), near the present Israeli port of Eilat.

Regardless, the name of this gin unequivocally recalls the spice trade, a trade that connected Italy with the rest of the world, from Morocco to Turkey, Spain to India, and even with mighty England.

The search for spices was a journey that began in the Malaysian islands, specifically the island of Malacca, where cubeb berries come from.

Joanne Moore has described this gin as follows: "I was inspired by the ways of the East and the various countries along the ancient spice route".

G&T GARNISH Sorrento lemon zest, ginger slice, grapefruit peel.

COCKTAILS Gin Toddy, Spicy Fifty Gin, Income Tax Cocktail.

OXLEY COLD DISTILLED GIN
47% ABV

WEB www.oxleygin.com

STYLE London Dry Gin

ORIGIN Clapham – United Kingdom

BOTANICALS 14. Juniper, grapefruit, lemon and orange peel, meadowsweet, vanilla, anise, orris root, liquorice, cocoa, grains of paradise, cassia, nutmeg, coriander.

PRODUCTION Instead of heating the vegetal elements and neutral grain spirit to a boiling point of approximately 78 ° C, here it is a cold distillation that uses vacuum pressure to lower the boiling point to 25-40 ° C. This way, the botanicals remain intact while the spirit is distilled. Oxley takes this technique to its extreme, distilling at −5 ° C.

This sub-zero distillation method ensures that the botanical's delicate flavors and oily compounds are preserved, rather than compromised or lost entirely through heat.

The second peculiarity of Oxley's distillation method is that, as a result of this drastic heat reduction, no heads or tails are created in the process. The gin is so purified that it lacks methanol or other unwanted substances.

All the botanicals are added together, in a one-shot distillation.

The botanical maceration period to create a batch is 15 hours. The spirit is then distilled (typically around 5-6 hours) and bottled at 47% abv. Usually 5 distillations run per week, producing an average of 600 bottles per process.

Before being sent to the distillery, the botanicals are vacuum-sealed in bags, with the exact weight required for each distillation. Every bag is equivalent to approximately 1 kg of botanicals, prepared at Bacardi's headquarters.

NOTES Owned by Bacardi, who launched it in 2009, following 8 years of experiments.

Bacardi's master herbalist is Ivano Tonutti, who handles 250-300 tons of spices and herbs gathered from virtually every corner of the planet.

Some botanicals are used fresh, while others, such as vanilla, are used in a dried version. In the first case, the bags are frozen or cryo dried to ensure maximum freshness of the essential oils inside them and preserve them for distillation.

G&T GARNISH Lemon peel, white grapefruit peel, Makrut lime leaf.

COCKTAILS Naked Martini, The Twentieth Century Cocktail, Negroni.

PINCKNEY BEND AMERICAN GIN
46.5% ABV

WEB www.pinckneybend.com

STYLE Distilled Compound Gin

ORIGIN New Haven, Missouri – U.S.

BOTANICALS 9. Juniper, coriander, angelica, orris root, liquorice, lavender, different citrus peels.

PRODUCTION Launched on November 21, 2011 and produced at the Pinckney Bend Distillery using a grains alcoholic base. Some of the botanicals are added after a separate infusion, while others are suspended in the still for vapor to pass through them. Everything is then expertly blended to obtain a perfect balance.

NOTES The project was the idea of 3 people: Jerry Meyer (CEO and president), Tom Anderson (a distiller with a degree in biochemistry) and Ralph Haynes (marketing manager).

Pinckney Bend was the old name of today's New Haven, a seaport town some 83 km north of St. Louis. It's a place known both for its whiskeys and for its dangerous Missouri river waters. In the 19th century, as many as 5 steamboats were destroyed along this stretch of the Missouri, including the famous Spread Eagle.

RELEASE Hand Crafted Hibiscus American Gin: 46.5% abv. Made with juniper, hibiscus flowers, yarrow and cubeb berries.

Hand Crafted Navy Strength American Gin: 57% abv.

Hand Crafted Cask Finished Gin: 46.5% abv. Aged in second passage whiskey barrels.

Hand Crafted American Vodka: 41.5% abv.

G&T GARNISH Citron peel, grapefruit peel, lavender flowers.

COCKTAILS Gimlet, Gin Sling, Ramos Fizz.

PORTOBELLO ROAD No. 171 GIN
42% ABV

WEB www.portobelloroadgin.com

STYLE London Dry Gin

ORIGIN London – United Kingdom

BOTANICALS 9. Juniper, lemon and bitter orange peel, cassia bark, nutmeg, angelica root, iris, liquorice, coriander seeds.

PRODUCTION Produced at the Thames Distillery using the Tom Thumb still. First production was in 2012 in the very small copper pot still located in their museum and named "Copernicus the Second".

NOTES Dedicated to the same-named London street the Ginstitute (a museum on the history of gin) is also located, where visitors can nose various botanicals and even prepare their own gin. The museum is located right above the Portobello Star Bar, at number 171 Portobello Road.

RELEASE Portobello Road Gin Local Heroes N.002: 42% abv. Created for the Italian market with the assistance of chef Carlo Cracco and Filippo Sisti, with the apparent addition of mango peel, Timut pepper and dill.

"A Parsnip in a Pear Tree" 4th ed.: 47.3% abv. Limited Edition Director's Cut No. 4, made with honey-roasted parsnip roots, different qualities of juniper, and other selected traditional botanicals. After distillation, pear liqueur is added. All raw materials are sourced from the famous Portobello market. Production is only 1,000 bottles.

Portobello Road Gin Pechuga N.003: 47.3% abv. Produced in perfect mezcal style (although in original Pechuga, distillation takes place after inserting the chicken or turkey breast into the still). Here, apples, pears, plums, currants, raisins, apricots, brown rice, passion fruit, cinnamon and cassia bark are also included, as well as mace and nutmeg. The turkey breast comes from Lidgate, a famous West London butcher.

Portobello Road Gin Smoky: 42% abv. Distilled in small batches on occasion of the 350th anniversary of the 1666 Great Fire of London. Features Irish peat, smoked juniper, Lapsang Souchong tea, chipotle and all the other traditional Portobello botanicals.

Portobello Road Navy Strength Gin: 57.1% abv. Made with English sea salt.

G&T GARNISH Garda lemon peel, thin cucumber peel, pink grapefruit peel.

COCKTAILS Monkey Gland, Negroni, London Mule.

GIN PRIMO
43% ABV

WEB www.ginprimo.com

STYLE Italian Distilled Gin

ORIGIN Cesena – Italy

BOTANICALS 5+1. Tuscan and Romagna juniper, santolina cenere, lemon verbena, lavender, Cervia sea salt.

PRODUCTION Designed and owned by Federico Lugaresi, this gin is produced in Asti at a recognized, quality distillery. The botanicals are strictly sourced from crops in the Romagna region, with abundant additions of Umbrian-Tuscan juniper mixed with the Spanish Juniper. The idea was born in 2015, and the gin was officially released in 2016 after a year of experiments. Infusions are made separately and the subsequent distillation is carried out in a bain-marie method in small pot stills. Next comes the assembly and the lowering of alcohol content, using the purest water from Romagna, with a perfectly balanced addition of Cervia sea salt.

NOTES The saline solution used for dilution corresponds to 1.891 grams per liter.

The old-fashioned label is truly *d'antan*: an interpretive watercolor portrait of Federico made by a painter from Faenza. Monica Zani.

Italian peasants used to name the firstborn male Primo.

G&T GARNISH Datterino tomato, lemon peel, sprig of lavender.

COCKTAILS Dirty Martini, Red Snapper, Gin Smash.

SABATINI GIN
41.3% ABV

WEB www.sabatinigin.com

STYLE London Dry Gin

ORIGIN Teccognano (Arezzo) – Italy

BOTANICALS 9. Tuscan juniper, Florentine iris, coriander, wild fennel, verbena, thyme, sage, lavender, olive leaves.

PRODUCTION Production and processing of botanicals take place in Villa Ugo in Teccognano, overseen by the Sabatini family. Only the juniper is harvested in several areas of the Val di Chiana (Arezzo). The distillation takes place in London at the Thames Distillery.

All botanicals are macerated together in pure grain alcohol for one night. A discontinuous distillation process follows using a 500-liter pot still.

NOTES From great-grandfather Guglielmo Giacosa, the family inherited a love for good wine and for the land. Giacosa was a 19th-century oenologist who exported Made in Italy products overseas, followed by some important vermouth productions in Chile and Australia. He also founded a distillery in his name, and some of his brands are still available on the market today (an egg-based marsala called Bornia for example, in Australia).

The 500-liter steel pot stills in which Sabatini Gin is produced are called Thumb and Thumbelina (in Italian, Pollicino and Pollicina). Both of these jewels were crafted by John Dore & Co, the historic company that created the first Carter Head alembic still.

In 2016, motivated by a desire to make their product available to admirers and professionals, the Sabatini family opened the gates of their estate to the public, organizing visits to the botanical garden and tastings of their smooth, blended product. The event took place in the renovated cocktail bar inside Villa Ugo's old *limonaia* (on Italian estates, the area where lemon trees are housed).

The alcohol content (41.3% abv) of Sabatini Gin is perfect for a juniper-based distillate, in which the botanicals balance the alcoholic strength. Yet another fortunate component is at work here, too: the sum of 4 + 1 + 3, equal to 8, is in Chinese culture (ba) considered the luckiest number of all. Moreover, the spelling recalls the infinity symbol, promoting prosperity and wealth.

G&T GARNISH Verbena leaves, fresh lavender sprig, lemon peel, fresh thyme sprig.

COCKTAILS Dirty Martini, Mediterranean Julep, Gin Fizz.

SACRED GIN
40% ABV

WEB www.sacredgin.com

STYLE Distilled Compound Gin

ORIGIN Highgate, London – United Kingdom

BOTANICALS 12, including juniper, angelica, cardamom, lemon zest, nutmeg, frankincense and liquorice.

PRODUCTION Founded in 2009 and distilled at the Sacred Microdistillery where, unlike conventional distilleries, the distillation takes place "vacuum" style in a glass still. Each batch produces around 250 bottles. Founder and master distiller Ian Hart has decided to modify his establishment by making substantial and significant technical changes that have improved the instrument used for his distillation. Each botanical is distilled separately starting with an English grain alcohol base.

NOTES The name of this gin derives from one of the botanicals used in its production: frankincense, from the genus Boswellia, including Boswellia *Sacra*.

Hart initially worked each individual botanical in 3 different distillation stages, or fractions: the first under the glass coil cooled with water at 0 ° C, the intermediate at −89 ° C cooled with dry ice, and the final using a liquid nitrogen coil at −196 ° C. Hart has also made improvements to the machine, removing the intermediate fraction (the dry ice).

RELEASE Sacred Organic Gin: 43.8% abv. Fully organic version with a higher percentage of juniper.

Sacred Old Tom Gin: 48% abv. Naturally sweetened, with liquorice and sweet orange notes.

Sacred Christmas Pudding Gin: 40% abv. Made according to a Victorian-era recipe for traditional English Christmas pudding, discovered in Hart's great-aunt's cookbook.

Sacred Organic Sloe Gin: 28.8% abv. Made with organic wild blackthorns left to macerate in a gin base for two and a half years.

G&T GARNISH Garda lemon peel, fresh thyme sprig, pink grapefruit peel.

COCKTAILS Cocktail Martini, Silver Fizz, Cardinale.

SIDERIT GIN
43% ABV

WEB www.destileriasiderit.com

STYLE London Dry Gin

ORIGIN Cantabria – Spain

BOTANICALS 12 stated, including juniper, mandarin peel, jamaica flowers (hibiscus), cinnamon, coriander, cardamom, bitter orange, pink allspice, rock tea (Sideritis Hyssopifolia), raw almond and lily root.

PRODUCTION Produced for the first time on February 1, 2013, after 131 tests using different formulas. The base distillate is of pure rye. All botanicals are macerated, then the product undergoes a double distillation in a very special still, one made of glass. According to the producers, each distillation using this sophisticated tool is equivalent to 5 distillations with a traditional alembic still.

NOTES The Latin name *Sideritis Hyssopifoglia* (a plant that also goes by "mountain tea" or "shepherd's tea") derives from the Greek word *sideros* (iron), on account of this plant's use in healing wounds caused by iron weapons. The first botanist to reference this name in modern times was Joseph Pitton de Tournefort (1656 – 1708).

The scientific nomenclature of this genus is owed to Linnaeus (1707 – 1778), the Swedish biologist who wrote *Species Plantarum*, and the father of our scientific classification system of living organisms.

Sideritis is known beyond Italy's borders as "malotira", a name dating to the period of Venetian rule over the island of Crete. The word is composed of two parts: "malo", meaning evil/disorder/disease, and "tira", meaning to pull out, an allusion to the beneficial properties of the plant. In addition to being a pleasant remedy for sore throats and respiratory system diseases, Sideritis's soothing properties make it an excellent cure for digestive tract disorders.

RELEASE Siderit Sherry Cask Gin Reserve: 43% abv. Aged 42 months in sherry barriques.

Siderit GingerLime Gin: 43% abv.

Siderit Hibiscus Gin: 43% abv.

G&T GARNISH Lemon peel, chamomile flowers, karkadè leaves.

COCKTAILS Abbey Cocktail, Limmer's Gin Punch, White Lady.

SIEGFRIED RHEINLAND DRY GIN
41% ABV

WEB www.siegfried.com

STYLE Distilled Compound Gin

ORIGIN Bonn, Rhineland – Germany

BOTANICALS 18 stated, including lime blossom, lavender, bitter orange, ginger and thyme.

PRODUCTION Characterized by an infusion of lime flowers added after distillation in a copper alembic. A 500-liter batch.

NOTES Produced in Germany by Gerald Koenen and Raphael Vollman's Rheinland Distillers, this gin was the most awarded gin in the world in 2015.

The name is inspired by Siegfried, hero of the Nibelungen saga who slayed the dragon and bathed in its blood to achieve immortality. However, a linden leaf landed on his back, leaving a deadly spot uncovered. The tale is strongly associated with Siegfried Gin, which is produced a stone's throw from the place where, according to legend, these events took place. And, as it happens, this gin's primary ingredient is from linden tree leaves.

G&T GARNISH Thin slice of ginger, Amalfi lemon peel, sprig of lavender.

COCKTAILS Last Word, Gin Julep, Gin Sling.

SIPSMITH DRY GIN
41.6% ABV

WEB www.sipsmith.com

STYLE London Dry Gin

ORIGIN London – United Kingdom

BOTANICALS 10. Juniper from Macedonia, bitter almond from Spain, coriander seeds from Bulgaria, cassia bark from China, liquorice root from Spain, angelica root from France, cinnamon from Madagascar, sweet oranges and lemons from Spain, iris root from Italy.

PRODUCTION Sipsmith Gin is distilled using 3 different stills: Prudence (300 liters) the oldest at the distillery and produced by Christian Carl in 1869, then Patience (300 liters) and Constance (1,500 liters). Cygnet (50 liters) is instead used for tests.

Sipsmith Gin is one-shot distilled, beginning with a neutral barley malt distillate. For dilution, water from Lydwell Spring, source of the Thames in Gloucestershire is used.

NOTES In 2009, Sipsmith was London's first distillery to use a copper pot still in 189 years.

Made by Sam Galsworthy and Fairfax Hall, joined by the great expert Jared Brown.

The name (sip + smith) refers to the small batch and entirely artisanal methods used here.

The original garage that was converted into the distillery was once the office of legendary beer and whiskey writer, Michael Jackson.

RELEASE Sipsmith V.J.O.P.: 57.7% abv. The acronym stands for Very Junipery Over Proof. Launched in 2003, this gin uses 75% more juniper than the classic London Dry Gin, and this botanical is added in 3 different moments during distillation.

Sipsmith Sloe Gin: 29% abv. Made with Dartmoor plums and the distillery's London Dry Gin.

Sipsmith Raffles 1915: 43% abv. Created on occasion of the 100th birthday of the Singapore Sling cocktail. The botanical line is inspired by Malaysia, and features ingredients such as jasmine flowers, pomelo peel, lemongrass, Makrut lime leaves, nutmeg and cardamom.

Sipsmith House of Commons Gin: 40.7% abv. Produced in 2015 for the House of Commons of the British Parliament.

Sipsmith Lemon Drizzle Gin: 40.4% abv.

G&T GARNISH Sorrento lemon peel, lime zest, fresh chamomile flowers.

COCKTAILS Martini Cocktail, Last Word, French 75.

SIX O'CLOCK GIN
43% ABV

WEB www.6oclockgin.com

STYLE London Dry Gin

ORIGIN Ashville Park, Short Way, Thornbury – United Kingdom

BOTANICALS 7. Juniper, orange peel, elderflower (locally grown), coriander, angelica, iris, winter savory.

PRODUCTION Produced by Bramley & Gage Ltd Artisan Spirit, a company founded by Edward and Penny Kain, who for over 30 years have been producing artisanal liqueurs of the highest quality. They were later joined by their children Michael and Felicity. This gin is handcrafted in very small batches using a small batch alembic still made by Arnold Holstein and named "Kathleen". It is a double sphere copper pot still, which increases contact with copper and therefore lends the distillate a more elegant note. In this style, botanicals are added later, to avoid a too-prolonged contact with alcohol and water. About 1,000 bottles at 43% abv result from each batch. The water used to lower the alcohol content is from Tarka Springs in North Devon.

NOTES The name alludes to the long-established family habit, enjoying a lovely Gin & Tonic every evening at 6 pm sharp.

The first gin they produced, that is the first Six O'Clock sample made, was not intended for sale but rather for use in the Organic Sloe Gin.

The distillers are said to have inherited the art of liqueur production from their parents/grandparents Edward Bramley Kain and Penelope Gage, fruit producers and retailers who often found themselves with all kinds of surplus product. To avoid waste, they started making liqueurs with the extras.

Charles Maxwell of Thames Distillery was called in to advise on the preparation and production of the recipe. In all, it took 31 months to develop this gin.

RELEASE Organic Sloe Gin: 26% abv.

G&T GARNISH Lemon peel, bergamot peel, sprig of yerba buena.

COCKTAILS Paradise, John Collins, Bijou.

GIN SUL
43% ABV

WEB www.gin-sul.de

STYLE Distilled Compound Gin

ORIGIN Altona, Hamburg – Germany

BOTANICALS Not stated. Juniper, coriander, lemons from Portugal, rosemary, chillies, lavender, cinnamon and rockrose resin.

PRODUCTION Distilled by the Altona Spirit Manufacture. The water used to lower the alcohol content comes from the Lüneburger Heide area, where there are 33 wells up to 326 meters deep that provide glacial water. Batches from an alembic still of only 100 liters.

NOTES Botanicals used in the preparation come from the south of Portugal.

Hand-bottled in exclusive white-glazed, screen-printed ceramic bottles.

Founder and master distiller Stephan Garbe started the project in 2013, dedicating it to the Portuguese land, its charm, scents, lights and that sensation known as *saudade*, the nostalgic state of longing felt when you leave.

RELEASE 2014: Ruby Sul: 46% abv. Aged in barrels that for 13 years prior contained Ruby port.

2015: Cruzeiro do Sul: 46% abv. Only 2,000 bottles. Aged in barriques previously used for Moscatel de Setúbal, a Portuguese fortified wine, with characteristic apricot flavors.

2016: Rota do Sul: 50% abv. Only 2,000 bottles. Featuring freshly picked thyme flowers distilled directly on Odeceixe beach. The recipe also includes orange, oregano, apricot and lemon peels.

2017: Kleine Freiheit: 45% abv. Only 4,000 bottles for this gin, named for one of the two most famous streets in the St. Pauli district of Hamburg (Kleine Freiheit means "little freedom"), once home to 17th-century artisans who were granted special religious and commercial freedoms. In this gin, juniper meets cumin, anise, fennel, dill, lemon and five types of pepper.

2018: Beijinho do Sul: 45% abv. Only 4,000 bottles. Bars in the historic center of Lisbon serve a dark red sour cherry liqueur called "Ginja". This gin is based precisely on just such a delicious base, of sour cherries macerated in a mixture of alcohol just a few hours after being harvested. The recipe also features Ceylon cinnamon, cloves and star anise.

G&T GARNISH Fresh rosemary, Garda lemon peel, fresh lavender.

COCKTAILS Bronx, Gin Sour, Gin Sling.

SYLVIUS GIN
45% ABV

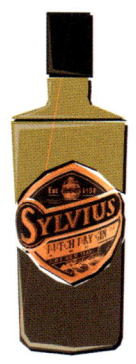

WEB www.sylviusgin.com

STYLE London Dry Gin

ORIGIN Schiedam – The Netherlands

BOTANICALS 10. Juniper, liquorice, cinnamon, star anise, lavender, cumin, angelica, coriander seeds, lemon and orange peel (in addition to the pulp).

PRODUCTION Produced in small batches by the Onder de Boompjes Distillery, founded in 1658. Each batch calls for approximately 30 kg of hand-cut oranges and 10 kg of lemons. Everything is macerated in Dutch grain alcohol, then subsequently distilled.

NOTES In honor of Dr. Sylvius, who in 1650 in Leiden was the first to produce a distillate based on juniper, called "Genievre".

The Boompjes Distillery was born in 1658 in Schiedam. It is one of the few surviving among the more than 180 distilleries once present.

G&T GARNISH Star anise seed, Garda lemon peel, lavender tuft.

COCKTAILS Martini Cocktail, Satan's Whiskers, Gin Sling.

THE BOTANIST GIN
46% ABV

WEB www.thebotanist.com

STYLE London Dry Gin

ORIGIN Isle of Islay, Argyll – United Kingdom

BOTANICALS 9 classic + 22 local.

The classics: local juniper, angelica root, cassia bark, cinnamon bark, coriander seeds, lemon and orange peel, liquorice root and orris root.

The locals: variegated mint, sweet chamomile, field thistle, hairy birch, elder flowers, spiny broom, hawthorn, heather, local juniper, yellow bedstraw, lemon balm, meadowsweet, mugwort, meadow red clover, spearmint, cicely, black nightshade, tansy, wild water mint, white clover, wild thyme, woodland germander.

PRODUCTION Produced at the historic Bruichladdich Distillery (founded in 1881) according to artisanal methods. Most of the 22 botanicals foraged among the hills and marshes surrounding the distillery. The botanicals are placed inside a low pressure Lomond pot still, affectionately baptized "Ugly Betty", and then distilled following a process that is about 3 times longer than the traditional process for gin (this is due to the low pressure processing). According to Jim McEwan, Bruichladdich's master distiller, distillation takes about 17 hours to complete. Inside the still is a basket specially designed for this type of production.

NOTES The still was designed by chemical engineer Alistair Cunningham and designer Arthur Warren in 1955 to create a variety of whiskey styles.

The key to its design was the thick column neck, which could hold 3 extra removable sections inserted for flexibility, emulating the effect of the different lengths of the "neck". One section housed 3 straightening plates that increased or decreased the reflux action. The plates could be opened at various angles from a horizontal to a vertical position.

First production was in 2010.

G&T GARNISH Amalfi lemon peel, fresh chamomile flowers, lemon balm flowers or leaves.

COCKTAILS Herbal Gin Smash, Foraged Gin Julep, Gimlet.

THE KING OF SOHO GIN
42% ABV

WEB www.kingofsohodrinks.com

STYLE London Dry Gin

ORIGIN London – United Kingdom

BOTANICALS 12 stated, including juniper, coriander, citrus (mostly grapefruit peel), angelica root, and cassia bark.

PRODUCTION Produced by the Thames Distillery, overseen by Charles Maxwell in two small copper pot stills, this is a traditional London Dry Gin. Distilled 4 times.

NOTES The name was chosen in homage to Paul Raymond, a former "porn baron" who went by the nickname "The King of Soho". This gin is his son Howard Raymond's creation, launched October 1, 2013.

Paul Raymond, who died in 2008, was a highly successful entrepreneur who saw his fortune increase by opening the first strip clubs in England (1958), and subsequently investing heavily in real estate, thereby earning his title. Today he is still very much remembered for his creativity, his business acumen and his "liberal" vision.

Paul Raymond partnered with Alex Robson, creating West End Drinks, owner of the brand.

The figures drawn on the label: a fox, animal of great cunning and creature of the night; a trumpet, meant to evoke the history of jazz in Soho; and a book to honor the publishing house founded by Paul Raymond.

RELEASE The King of Soho Gin Variorum: 37.5% abv. Pink Berry Edition of the original, with strawberry and chamomile flowers.

G&T GARNISH Lemon peel, grapefruit peel, liquorice stick.

COCKTAILS Lord Chamberlain, Seventh Heaven, John Collins.

THE LONDON N.1 ORIGINAL BLUE GIN
47% ABV

WEB www.thelondonn1.com

STYLE Distilled Gin

ORIGIN Jerez de la Frontera – Spain

BOTANICALS 12. Juniper, almond, angelica, bergamot, orange and lemon peel, cinnamon, cassia, coriander, liquorice, iris, winter savory.

PRODUCTION Owned by the Spanish giant Gonzàlez Byass (founded in 1835) yet produced in London at the Thames Distillery, where the 3 distillations take place exclusively in the pot still system. The starting neutral alcohol is of the highest quality Suffolk and Norfolk wheat. Before being put on the market, between maceration and resting it passes at least 3 weeks in stainless steel tanks.

NOTES The color derives from the final infusion of gardenia flowers and bergamot oil in the distillate.

The name "gardenia" was attributed by the naturalist John Ellis (1710 – 1776) in honor of the naturalist Alexander Garden (1730 – 1791), a North American of Scottish origin who was a known friend of Linnaeus.

In 1920, the popularity of the gardenia grew exponentially, as it was associated with the figure of the viveur, a type of gentleman who frequented elegant, sophisticated evening parties, and who would pin this white, fragrant flower in his buttonhole.

In the language of flowers and plants, gardenia is the symbol of kindness and sincere friendship, but also of fleeting beauty. Today it is also a symbol of solidarity in the fight against multiple sclerosis.

G&T GARNISH Sorrento lemon peel, bergamot peel, gardenia flowers.

COCKTAILS London Mule, Gin Cobbler, Gin Sour.

UNCLE VAL'S BOTANICAL'S GIN
45% ABV

WEB www.unclevalsgin.com

STYLE American Cold Compound Gin

ORIGIN California, but also produced in Oregon – U.S.

BOTANICALS 5. Juniper, lime, sage, lavender, cucumber.

PRODUCTION Produced by 35 Maple Street Distillery, it was launched in March 2012. All the herbs used are gathered in Tuscany, in gardens of Lucca, by uncle Valerio (Valerio Cecchetti, retired doctor), to whom the gin is dedicated. The neutral distillate base is obtained from wheat, carried out in 5 steps in pot stills with 100% natural botanicals. The spices are macerated in alcohol for several hours inside an immersion bag (like a large tea bag). The result is then filtered 5 times using coals and pumice, lowering the impurities to 0.0005%.

The lowering of alcohol content uses fresh water from the Cascade Mountains. Hand-labelled bottles.

NOTES In 2015, The Other Guys decided to change their name to "3 Badges" after moving their administrative offices to a former fire station in Sonoma, California. Founder August Sebastiani, a winemaker with Tuscan origins (Valerio Cecchetti is a relative) emigrated in 1895 and founded the company. He discovered numerous documents that testified to his family's participation in the construction of the building. His grandfather had worked as a volunteer and received 3 badges for his career (thus "3 Badges" symbolizes returning to one's roots and honoring passion and craftsmanship as the cornerstone of the company). In 1904, Sebastiani then bought enough land in Sonoma to open his own cellar. His was the only local company to remain open during Prohibition.

RELEASE Uncle Val's Peppered Gin: 45% abv. Based on chilli, juniper, black pepper and pimento.

Uncle Val's Restorative Gin: 45% abv. Infused with cucumber, coriander, rose petals and juniper. Distilled 5 times.

G&T GARNISH Sage leaves, lime peel, lavender sprig.

COCKTAILS Paradise, American John Collins, Gin Cobbler.

UPPERCUT PREMIUM QUALITY GIN
49.6% ABV

WEB www.spring-gin.be

STYLE London Dry Gin

ORIGIN Baarle-Nassau – The Netherlands

BOTANICALS 7. Juniper, damiana leaves, dandelion, nettle, strawberry leaves, liquorice, verbena.

PRODUCTION To create this gin, Manuel Wouters teamed up with Zuidam Distillers in Baarle-Nassau, one of the last independent artisanal distilleries in the Netherlands, selecting 4 herbs famous in the past for keeping men fit and healthy. In addition to being a remedy for headaches, damiana is also associated with curing some minor sexual problems; while strawberry leaves promote digestion, liquorice is good for the throat, and verbena is a precursor to our modern toothpaste. The still used is a classic copper still.

RELEASE Left Hook Gin: 47.2% abv. With juniper, pink pepper, tarragon, ginger, mace and bay leaves.

G&T GARNISH Sage leaf, verbena bunch, Garda lemon peel.

COCKTAILS Gimlet, Corpse Reviver No. 2, Gin Cobbler.

VL92
41.7% ABV

WEB www.vl92.com

STYLE Distilled Gin

ORIGIN Vlaardingen – The Netherlands

BOTANICALS Not stated, but definitely juniper, coriander leaves, orange blossoms.

PRODUCTION Produced by Van Toor Distilleerderij starting with a compound of 25% malt wine (the base of a genever).

NOTES Launched by two friends, Leo Fontijne and Sietze Kalkwijk, who produced the first batch on May 15, 2012.

Named after a historic Dutch sailing vessel that transported the exotic spices from the Indies.

The packaging is inspired by a bottle of saline solution.

The Dutch shipped the first genever to the United Kingdom in the early 17th century. And on May 15, 2012, the first overseas shipment of VL92 was sent to London. The gin arrived in the British capital on a historic Dutch ship, having set sail from its port in Vlaardingen. Sietze Kalkwijk presented the first VL92 package to the Artesian Bar team, headed by famous barman Alex Kratena.

RELEASE VL92 YY: 45% abv. Contains 55% malt wine aged in wood, like a genever. Unfiltered to preserve its aroma. Limited series.

VL92 1736 Gin Act XY: 45% abv. Limited edition. A tribute to the well-known Gin Act of 1736, this gin is produced from 100% malt wine genever by Van Toor, to which some classic 18th-century botanicals are added, such as juniper (naturally), angelica, iris root, coriander seeds, and violet in the final stage of distillation.

G&T GARNISH Slice of ginger, lemon peel, oyster leaf.

COCKTAILS Bijou, Hanky Panky, Gin Cocktail (Harry Johnson).

VOR ICELANDIC GIN
47% ABV

WEB www.vorgin.is

STYLE London Dry Gin

ORIGIN Garðabær – Iceland

BOTANICALS 9. Juniper, rhubarb root, angelica, birch leaves, wild thyme, black crowberry, Icelandic lichen, sweet seaweed, kale.

PRODUCTION Produced by Eimverk Distillery, which was founded in 2009 on the outskirts of Reykjavik, where the family of the same name chose to produce the first 100% Icelandic distillate, starting from their own organic barley. Their dream came true in 2014. Once the mash is produced, they distill it twice in a pot still, to create a new-make spirit in perfect whiskey style. At this point, they produce the well-known whiskey called "Floki", or they use the same base for their botanical cocktail, made solely from organic ingredients, and redistill everything. Botanicals are left to soak in the spirit for about a week, and are then redistilled all together in a 300-liter Arnold Holstein.

NOTES "Vor" means "spring" in Icelandic, and in fact the Arctic spring is the inspiration for this gin. It is produced in small batches of 500 bottles each.

A small but curious observation on this project: Eimverk Distillery produces the base distillate with which it also produces whiskey, at a concentration between 67% and 80% abv—an astounding concentration if the product were entirely destined for whiskey production. Regarding gin, however, EU guidelines require instead that the alcoholic concentration of a base distillate must have a minimum alcohol content of 96% abv. This is namely to strip the base spirit almost completely of any aromatic component, rendering it neutral in all respects. One could therefore almost say that here, in a certain sense, "the law is broken".

RELEASE Vor Barrel Aged Reserve Icelandic Gin: 47% abv. Launched in 2015. Aged in whiskey oak barrels for 6 weeks.

Vor Sloe Icelandic Gin: 21% abv. Uses the base of their oak aged gin with the addition of blueberries, black crowberry, and sugar in maceration.

Vor Navy Strength Icelandic Gin: 57% abv.

G&T GARNISH Lemon peel, sprig of thyme or rosemary.

COCKTAILS Gin Julep, Gin Old Fashioned, Martinez, John Collins.

WARNER EDWARDS HARRINGHTON DRY GIN
44% ABV

WEB www.warnersdistillery.com

STYLE London Dry Gin

ORIGIN Harrinhgton, Northamptonshire – United Kingdom

BOTANICALS 10 + 1 secret. Cardamom from Guatemala, angelica root from Holland, black pepper from Vietnam, cinnamon from Sri Lanka, local elderberry, Italian juniper, Spanish orange, coriander, nutmeg, lemon zest.

PRODUCTION The still created by the now legendary Arnold Holstein is completely copper and includes 8 copper plates (they only use 4) and a copper catalyser called "Curiosity". The base distillate is purchased from Langleys. They do not macerate in alcohol, but rather use a direct distillation of the botanicals. Essentially, Curiosity is filled with water and alcohol and the recipe (perfectly measured amounts of botanicals) is poured into it. Once the alembic is closed, distillation is immediately carried out in one-shot style, an added value for this precious gin. The process takes about 12 hours, and is followed by 16 days of rest in steel tanks, then bottling and hand labeling.

NOTES Initially, partners Tom Warner and Tina Warner-Keogh wanted to create an essential oils business, focusing on lavender. But when they discovered that lavender is a very difficult flower to manage (in terms of freshness), they thought about the distillation of vodka, self-producing the grain in their own farm fields. However, the capital investment needed to fulfill this plan would have been too risky for them. For this reason, they decided to stay at the Falls Farm with Tom's family, where the spring from which they draw water to lower the alcohol content of their spirits is also found. The distillery was inaugurated in December 2012.

RELEASE Warners Rhubarb Gin: 40% abv.
Warner's Edelflower Gin: 40% abv.
Warner's Lemon Balm Gin: 43% abv.
Warner's HoneyBee Gin: 43% abv.
Warner's Sloe Gin: 30% abv.

G&T GARNISH Garda lemon peel, yellow grapefruit peel, lavender tuft.

COCKTAILS Income Tax Cocktail, John Collins, Negroni.

WHITLEY NEILL DRY GIN
42% ABV

WEB www.whitleyneill.com

STYLE London Dry Gin

ORIGIN London, but distilled in Birmingham – United Kingdom

BOTANICALS 9. Juniper from Italy (Umbria and Tuscany) and India, baobab fruit, alkekengi (bladder cherry) from Africa, sweet orange and lemon peel from Spain, angelica root from Western Europe, Florentine iris from Italy, cassia bark from Southeast Asia.

PRODUCTION Owned by Whitley Neill Limited, founded by Johnny Neill (born in 1972) in May 2004 to fulfill his dream of creating new, interesting and innovative Premium drinks. Neill, a descendant of a family of distillers whose tradition dates back as far as 1762, was born less than 1 km from the family distillery in Warrington, England.

This is the only gin in the world that features two never-before-used African botanicals. Distilled at Langleys Distillery in Birmingham, the grain used to make the base has a purity of up to 96%. The distillation system is discontinuous and takes place in a 100-year-old alembic pot still named "Constance". The gin was launched in 2005.

NOTES The label depicts a stylized baobab tree (also called the "tree of life"). The company donates 5% of its proceeds to Tree-Aid, a charity that helps the poorest African families.

Baobabs are deciduous trees that reach heights between 5 and 25 m. The diameter of the trunk, on the other hand, varies from 7 to 11 m (in truly exceptional cases). This tree is famous for its ability to store water (up to 120,000 liters) inside its trunk, thus able to withstand the harsh drought conditions in the areas where it is widespread. Pollination occurs through a particular species of bats and a type of nocturnal lemur.

The name "baobab" derives from the Arabic *buhibab*, meaning "fruit with many seeds".

Whitley Neill's slogan is "Made in London and inspired by Africa".

RELEASE Blood Orange Gin: 43% abv. With Sicilian orange.
Raspberry Gin: 43% abv.
Rhubarb & Ginger Gin: 43% abv.
Quince Gin: 43% abv. Made with quince.
Parma Violet Gin: 43% abv. Infused with violet flowers.

G&T GARNISH Lemon peel, alkekengi (bladder cherry), red berries.

COCKTAILS Gin Sling, Fruit Gin Sour, Angel Face.

WILD WOMBAT GIN
42% ABV

WEB www.wildwombatspirits.com

STYLE Australian Distilled Gin

ORIGIN Australia

BOTANICALS 11 stated. Juniper, lemon myrtle, apple, cassia, orange, coriander seeds, angelica root, cinnamon, cloves, nutmeg, cardamom.

PRODUCTION The base distillate is made from Australian wheat distilled in a multiple plate column still, then diluted with rainwater. That's right! They use rainwater to dilute their distillate, in which the 11 different botanicals, obviously of Australian origin, will then be infused, macerated and redistilled separately in a small batch copper still, then be combined into a single blend.

NOTES The creator and owner of the brand is NASC (New Australia Spirits Company Group), the first Australian company to have designed both a distillery and a brand from scratch.

The wombat, belonging to the Vombatidae family, is a marsupial that lives in Australia, where it has been nicknamed "Mr. W". Physically, wombats resemble rodents, but with short, stubby legs and a small tail. The name wombat is owed to the Eora, an Aboriginal population who once lived in what is now the Sydney area. The wombat can reach a speed of 40 km per hour. The females have a curious way of "declaring themselves available" towards males: during their fertile period, they bite their partner's butt! Curious, too, that the pouch where females grow their young is on their backside rather than the belly.

RELEASE Wild Wombat Bush Legend I: 42% abv. With juniper, passion berry, lemon, lime, Davidsons plums, Australian mint and iris root.

Wild Wombat Bush Legend II: 42% abv. With juniper, quandong, kakadu, wattleseed, Australian Murray River salt and lime.

Wild Wombat Sloe Gin: 42% abv. With juniper and quandong from the Australian Outback.

Wild Wombat Jolly Roger Strength Gin: 58.8% abv. With juniper, orange, star anise, cardamom, cloves, kakadu, wattleseed and a natural black dye.

G&T GARNISH Lemon peel, thin apple slice, sprig of calamint.

COCKTAILS Gin Julep, SouthSide, Gin Old fashion.

WINDSPIEL PREMIUM DRY GIN
47% ABV

WEB www.windspiel-manufaktur.com

STYLE Distilled Compound Gin

ORIGIN Daun, Rhineland – Germany

BOTANICALS Among those stated, juniper, coriander, cinnamon, citrus peel, ginger and lavender.

PRODUCTION Master distiller Holger Borchers makes this Premium gin directly from a 100% Eifel potato distillate in his hometown in Northern Germany. After the fall harvest, the potatoes are then ground, mixed with water and heated, to start the process that will lead to sugar production. Finally, yeasts are added to ferment everything, then comes a triple distillation (two in a continuous and the third in a 150-liter still). Then the individual botanicals are immersed and distilled separately.

NOTES The three founders of the company, Sandra Wimmeler, Denis Lönnendonker and Tobias Schwoll, decided to move to this region in search of peace and tranquility, purchasing the Weilerhof farm in 2008. They grew elephant grass (or Napier grass) first, and then potatoes.

Tobias's dream was to grow potatoes from the nearby Eifel plateau. However, unlike the neighboring area, the land at his disposal was not suited to the tuber's production. It therefore took much effort and perseverance to make the vegetable in question grow successfully (considering, too, the stony area, the Vulkaneifel). The Eifel potato then became a central ingredient of their diet, along with a good glass of gin after dinner. Hence, the ultimate idea: a Premium Gin using this precious tuber.

The Windspiel label, which in German means greyhound, has its origins in a legend. According to inhabitants of the area, the first person to enjoy the Eifel potato was Frederick II of Prussia, in 1757. Given that the sovereign's other great passion was greyhounds, Sandra Wimmeler decided to feature that particular animal on their bottles.

The blue and white label bears two expressions: "Frederick the Great" and "Faithful to the command of potatoes".

RELEASE Windspiel Premium Sloe Gin: 33% abv.
Windspiel Premium Dry Gin Reserve: 49.3% abv.
Windspiel Premium Dry Gin Navy Strength: 57% abv.
Windspiel Barrel Aged Potato Vodka: 42% abv.

G&T GARNISH Sprig of lavender, yellow grapefruit peel, slice of ginger.

COCKTAILS Gin Crusta, French 75, Gin Gin Mule.

Index
Bibliography – Acknowledgments

Index

A

Almond	82
Ambrosia Wine Cobbler	71
Angelica	94
Angostura Bitter	23, 141–143
Aviation American Gin	192

B

Basil	88
Bathtub Gin	193
Beach, Donn	155, 156
Beefeater 24	171
Beer Street	21, 25–27
Belle Époque	153
Bennett	100
Bergamot	90
Berkeley Square Gin	194
Berry Tray Still	104
Bickens London Dry Gin	195
Big Gin	196
Bitter Orange	90
Blackwoods Vintage Dry Gin	197
Blue Ribbon Essential Gin	200
Bluecoat Gin	201
Bobby's Dry Gin	202
Bombay Sapphire Dry Gin	173
Bootlegger 21 Gin	203
Botanicals Chamber	103
Boxer Gin	204
Brecon Special Reserve Gin	205
Brockmans Premium Gin	206
Brooklyn Handcrafted Gin	207
Buddha's Hand	76
Bullards Norwich Dry Gin	208
Bulldog Gin	175
By the Dutch Gin	209

C

Camillo, Conte	148, 149
Canaïma Gin	210
Caorunn Gin	177
Cardamom	81
Cardinal Gin	211
Carter Head	103
Carter Head Still	101
Cassia	74
Ceccarelli, Giovanni	108
Chamomile	80
Cinnamon	85
Citadelle Gin	212
Clove	78
Coffey, Aeneas	22, 102, 103
Cold Compound Gin	120
Cold Distillation	111
Column Still	98
Copenhagen Dry Gin	213
Copperhead Gin	214
Coriander	86
Cotswolds Gin	215
Cubical Premium	216
Cucumber	87

D

de la Böe, Franciscus	16
de la Boue, Franciscus Sylvius	16
Death's Door Gin	217
Distilled Gin	119
Dodd's Gin	218
Dragon's Eye	74
Dry Fly Handcrafted Gin	219
Dudley, Robert	17

E

Eden.Mill St. Andrews Gin	220
Edinburgh Gin	221
Elderberry	83
Elephant London Dry Gin	222

F

Fennel	75
Ferdinand's Saar Dry Gin	223
Filliers 28 Gin	224
Flavoured Seaport	125
Fog Cutter	165
Forest Dry Gin	225
Four Pillars Rare Dry Gin	226
Fred Jerbis Gin43	227
Fynoderee Manx Dry Gin	228

G

G' Vine Gin	229
Gastro Gin&Jonnie	230
Gay-Lussac, Joseph	112
Geranium	75
Geranium Gin	231
Giass Dry Gin	232
Gimlet	23, 24, 32, 55
Gin Act	21, 28
Gin Craze	19–21, 46
Gin de Mahon	121
Gin del Professore	233
Gin Lane	21, 25, 26
Gin Mare	249
Gin Mezcal	121
Gin Palace	22, 30
Gin Primo	262
Gin Sul	269
Ginebra San Miguel Gin	123
Ginepraio Tuscan Dry Gin	179
Ginger	81
Grains of paradise	76
Granit Bavarian Gin	243
Green Hat Gin	235
Greenall's Bloom Gin	198

H

Half Hitch Distilled Gin	236
Hendrick's Gin	237
Herbal G&T	129
Hernö Gin	238
High proof	51
Hogarth, William	21, 25–27
Honeysuckle	77
Hops	83

I

Iris (Florentine)	94
Ish Gin	239

J

Jenever	19, 24, 37
Jensen's Bermondsey Dry Gin	240
Jinzu Gin	241
Juniper	92, 93
Junipero Gin	242

K

Kamaaina	161
Karst gin	121
Karst juniper brandy	121
Kelbo's Scorpion	167
Ki No Bi Kyoto Dry Gin	243
Kraški brinjevec stà	121

L

Lavender	73
Le Tribute Gin	244
Lemon	90
Lemon balm	85
Leopold's Gin	245
Leuci, Leonardo	7–9
Lime	91
Liquorice	77
London Dry Gin	56, 118
London Gin	118
Lotus flower	95
Luz Fashionable Gin	247

M

Macaronesian Gin	248
Makrut lime	91
Martin Miller's Gin	181
Mayfair Gin	250
Mayflower	50
Mediterranean Martini	63
Mint	87
Mombasa Club Dry Gin	251
Monkey 47 Schwarzwald Dry Gin	252

N

N°3 London Dry Gin	246
N°209 Gin	253
Navy Strength	51
Nolet's Dry Gin	254
Nordes Atlantic Galician Gin	255
Nutmeg	89

O

Old English Gin	256
Old Tom	44
Olive	78
O'ndina Gin	257
Opihr Oriental Spiced Gin	258
Orange	90
Origami	43
Orris root	94
Oxley Cold Distilled Gin	259

P

Panarea Island Gin	183
Pepper	84
Peychaud, Antoine Amédée	142
Picchi, Luca	148
Pilgrims	50
Pinckney Bend American Gin	260
Pink Gin	23, 31, 142
Plymouth	50
Pogo Stick	163
Pomelo	91
Portobello Road No. 171 Gin	261
Pot Still	98
Prohibition	29, 155

Q

Quiet Martini	61

R

Raspberry	82
Reisetbauer Blue Gin Vintage	199
Reverend Stoughton	31
Rivo Gin	185
Roby Marton Gin	187
Rose hip	79
Rosemary	74
Rotavapor	108
Royal Hawaiian	159

S

Sabatini Gin	263
Sacred Gin	264
Saffron	88
Sage	73
Scarselli, Fosco	148, 149
Schweppe, Johann Jacob	140
Siderit Gin	265
Siegert, Johann Gottlieb Benjamin	142, 143
Siegfried Rheinland Dry Gin	266
Sikes, Bartholomew	112
Sipsmith Dry Gin	267
Six O'Clock Gin	268
Sloe Gin	64
Sloe_Motion	67
Star anise	80
Steinhäger	121
Stoughton, Richard	141
Super Premium	216
Sylvius Gin	270

T

Tanqueray Ten Gin	189
Tea	89
The Botanist Gin	271
The King of Soho Gin	272
The London N.1 Original Blue Gin	273
Thyme	79
Tiki	154–157
Top Hat	49
Tudor, Frederic	137

U

Uncle Val's Botanical's Gin	274
"Unusual" Negroni	151
Uppercut Premium Quality Gin	275

V

Vacuum Distillation	108
Vic, Trader	155–157
Vilniaus Džinas	121
Vilnius Gin	121
VL 92	276
Vor Icelandic Gin	277

W

Waragi	123
Warner Edwards Harringhton Dry Gin	278
Whitley Neill Dry Gin	279
Wild Wombat Gin	280
Windspiel Premium Dry Gin	281
Winter savory	86

X

Xoringuer	121

Y

Yuzu	91

Z

Zottola, Gianni	154

BIBLIOGRAPHY

- Dave Arnold – *Liquid Intelligence* – Readrink
- Richard Barnett – *The Dedalus Book of Gin* – Dedalus Limited
- Thea Bennet – *London Gin: The Gin Craze* – Golden Guides Press Ltd
- Dave Broom – *Gin. The manual* – Octopus Publishing Group
- Jared Brown, Anistatia Miller – *Spirituous Journey: A History of Drink*, 2 volumes – Readrink
- Simon Difford – *Gin. The Bartender's Bible* – Firefly Books
- Patrick Dillon – *Gin, The Much-Lamented Death of Madam Geneva* – Thistle Publishing
- Frédéric Du Bois and Isabel Boons – *The Complete Guide Gin & Tonic for The Perfect Mix* – Lannoo
- Adam Elmegirab – *Dr. Adam Elmegirab's Book of Bitters* – Dog 'n' Bone
- Lesley Jacobs Solmonson – *Gin: A Global History* – The Edible Series, Reaktion Books
- Harry Johnson – *Bartenders' Manual* – Mud Puddle Books
- Aaron Knoll – *Gin. The Art and Craft of the Artisanal Revival in 300 Distillations* – Aurum Press Ltd
- Aaron J. Knoll e David T. Smith – *The Craft of Gin* – White Mule Press
- Fulvio Piccinino – *Il gin italiano* – Graphot
- Gary Regan – *The Bartender's Gin Compendium* – Xlibris
- Tristan Stephenson – *Gin Palace* – Royal Peters & Small
- Davide Terziotti – *Lo spirito del gin* – WhiteStar
- Jerry Thomas – *How to mix Drinks* – Cocktail Kingdom
- Jerry Thomas Project – *Twist on Classic. I grandi cocktail del Jerry Thomas Project* – Giunti
- David Wondrich – *Imbibe* – A Perigee Book

ACKNOWLEDGMENTS

There are truly so many people I would like to thank for this book. I've no idea how many gin and tonics I will have to offer you all, but I am sure it will be a mighty amount indeed. Nevertheless, my heartfelt thanks go to those of you I hold dear to my heart. My family, for starters: my mom Milena, my father Gianni, and my brother Edoardo. Then, my partner as well as the illustrator of this book, Serena, who has never asked me to change my mind, and has always given me the energy to face every challenge. And thanks to Maurizio, "the pen" who wished to accompany me on my "sentimental journey" into the world of gin.

Samuele Ambrosi

© Guido Tommasi Editore – Datanova S.r.l., 2021

Text: Samuele Ambrosi, Maurizio Maestrelli
Illustrations: Serena Conti
Graphics: Carolina Quaresima
Translation: Amy Gulick

Any reproduction, partial or total, on any device,
in particular photocopy and microfilm, is strictly prohibited
without the express written permission of the publisher.

ISBN: 978 88 6753 345 9

Printed in EU